D0095240

SKY
RANCH

SKY RANCH

Living on a Remote Ranch in Idaho

BOBBI PHELPS

Skyhorse Publishing

Skyhorse Publishing books may be purchased in bulk at special discounts for sales promotion, corporate gifts, fund-raising, or educational purposes. Special editions can also be created to specifications. For details, contact the Special Sales Department, Skyhorse Publishing, 307 West 36th Street, 11th Floor, New York, NY 10018 or info@skyhorsepublishing.com.

Skyhorse® and Skyhorse Publishing® are registered trademarks of Skyhorse Publishing, Inc.®, a Delaware corporation.

Visit our website at www.skyhorsepublishing.com.

10 9 8 7 6 5 4 3 2 1

Library of Congress Cataloging-in-Publication Data is available on file.

Cover design by Paul Qualcom and Bob Ballard.

Print ISBN: 978-1-5107-5107-1
Ebook ISBN: 978-1-5107-5109-5

Printed in the United States of America

Dedicated to
Georgina and Don Wolverton
Nancy and Doug Strand

Without them I could not have survived the rigors of rural life,
the challenges and confrontations of Sky Ranch.
They brought love and happiness to me
when I felt as if I couldn't go on.
They were my rocks.
My champions.

Table of Contents

Disclaimer

Sky Ranch reflects the recollections of my Idaho ranch experiences between 1980 and 1996: the adventures in a time before camera phones, GPS technology, and social media.

The conversations written on the following pages have been recreated to evoke the substance of what was said. Some names and identifying characteristics have been changed. Sky Ranch is the fictional name for the Wolverton's family ranch: Golden Valley Land and Cattle Company.

While all the incidents described in this book are true to the best of my memory, certain events have been compressed, consolidated, or rearranged to aid in the narrative flow. *Sky Ranch* is not intended to be an exact duplication but an effective representation of a city girl's time at a remote ranch in Idaho.

—Bobbi Phelps

Foreword

Sky Ranch by Bobbi Phelps is a series of kindly, authentic, and sensitive profiles of finding herself on a remote Southern Idaho ranch, in love with the cowboy-owner she later marries. As is said, "all is not sweetness and light."

If you're not from the West and not familiar with this part of the country, it might be tempting to see ranch life through an idyllic hazy glow of romanticism and fantasy. Phelps brings us to the place as it really was, and as it remains today.

This is a woman who—before her time in Idaho—had never ridden in a pickup truck, watched an animal die, touched (only once) an electric cattle fence, survived a howling ground blizzard, duck hunted, or bore and raised a son whom she and her husband nearly lost in a later farm incident.

Brought up in a tony suburb in Connecticut, Phelps graduated from UC Berkeley and had an early career in the airline industry. Phelps then took a job as an advertising sales representative for Southern Idaho's area newspaper in 1980. She was in her late thirties, but as the rookie, she was handed sales accounts no one else wanted, including a rendering plant that processed dead animals. (Yep, they advertised.) She was indeed a spunky woman, learning to live in and love Western ranch life.

Phelps writes with a fine eye for detail, place, and remembrance. She's particularly good at describing the people around her, miles away on gravel roads yet the closest neighbors. Her accounts are tight and brisk with the clipped definitive of Western common speech. Although she's no longer in Idaho, Phelps has retained our dialect and phrasing in both her memory and in the vignettes of *Sky Ranch*.

This is no fictionalized account, nor is it an exact memoir. *Sky Ranch* is populated with real people whom Phelps names and quotes. Though she says some conversations are reconstructed, they have the cadence and pace of overheard and yet remembered incidents.

As the Rockies were settled, there were many accounts, some written years later, of the challenges faced by those early pioneers. It's a genre of Western writing which has somewhat gone out of style but shouldn't have. Phelps's form is conversational and seemingly a bit wistful, as if she is relating how fortunate she was to have been there at that moment. *Sky Ranch* conveys a crisp late summer evening with a languid harvest moon and the golden wheat and barley shimmering in the gathering dark.

Her remembrances of ranch life are in and of the region. They'll remind readers of the genre of Western literature of *Lonesome Dove, O Pioneers!, Little House on the Prairie,* and *English Creek.* These are all set in a past time, but the Rocky Mountain life is still extant in regions like Southern Idaho's windswept high plains and among the hardy, resilient people who live there. Phelps's challenges seem no less real.

Sky Ranch puts Phelps clearly in two groups of Western life: those who have lived it, and those who have told its story well.

—Rep. Stephen Hartgen, ret.
Idaho House of Representatives

Preface

All I could see in any direction was sagebrush and prairie grass. In the far distance rose the mountains surrounding Golden Valley, a large section of South Central Idaho. It was August 1980. Somewhere on the endless expanse, Sky Ranch spread over the rough grassland, four thousand acres zigzagging across six by ten miles, almost the size of my hometown in the Northeast. Although I had traveled the globe as a flight attendant, this was a world I had never experienced. In my late thirties, I enjoyed life to its fullest. I gained information about farm and ranch life from a good-looking cowboy, a man who later became my husband.

Raised near New York City, my life appeared typical of those living in American suburbs. But it was not. My life was only typical of many of us living on Connecticut's "gold" coast, a land of movie stars and business executives. My mother stayed home and enjoyed the trappings of a successful husband. She played tennis and bridge, became president of our town's garden association, and chaired my school's annual fair. My father worked in New York City's real estate industry, belonged to the University Club in Manhattan, volunteered at the local fire department, and was a charter member of the Darien Country Club. He lived the classic, corporate lifestyle.

Before I started kindergarten, my mother enrolled me in piano, horseback riding, and gymnastics. During summers, my family vacationed on Cape Cod where we swam in the warm Atlantic

Ocean and collected colorful shells on white sandy beaches. My sisters and I attended camp in New Hampshire, beach parties on Long Island Sound, and debutante galas at elaborate homes throughout Fairfield County.

I was lucky to have two wildly different but loving parents. My father kept me somewhat grounded and my mother encouraged me to be curious, to challenge, and to explore. I identified more with my exuberant mother, oscillating between elation and fear. In my early twenties, my feelings of fear dissipated. I became an international flight attendant, taking troops in and out of Vietnam during the height of the war.

After completing six years in the airline industry, I retired and married my college sweetheart. Using previously purchased airline passes we traveled the world for eighteen months, camping and fishing. He was a writer for *Fly Fisherman Magazine* and I illustrated his articles with my photographs. When we returned to the States I began selling my fishing photographs to sport magazines. They were well received and I started the Angler's Calendar Company in 1975, the first fly-fishing calendar on the American market.

Following a divorce and my marriage to Mike Wolverton, an Idaho rancher, I discovered what it was like to live in rural America. New adventures awaited me as my life turned upside down. The contrast between my suburban background and my farming future created amazing differences, both challenging and rewarding. I observed and learned, ready to change just like the seasons. But I was a stranger in this incredibly strange land. There were no native trees, just prairie grass and sagebrush; no oceans, just mighty rivers; and no rain to speak of, only powerful winds and violent snowstorms. My Connecticut family lived two thousand miles to the east, and my ranch neighbors lived far away—down dirt roads that often became impassable from winter storms or spring runoffs. I felt like an interloper on a giant learning curve, facing each challenge with reckless innocence.

Chapter One

Meeting Cowboy Mike

"Are the boxes packed? Is there space for more?" I asked Alanna, my assistant at the Angler's Calendar Company.

"We're jammed full. Just enough room for Ronni and her suitcase."

"Here she is," I said as she parked next to my yellow Mini Cooper, a tiny station wagon with bucket seats in front and a bench seat behind.

"Good timing, Ronni. We're packed and ready to go. There's a small spot for you in the back."

She came closer, bent over, and peered in the car's window. Straightening, she brushed her blond bangs from her face and asked, "Where? There's no room."

"Sure, there is. Right behind Alanna. Not to worry," I said as I opened the passenger door and pushed the front seat forward. Ronni, one of my best friends from our airline days, wedged herself between cardboard boxes that had been crammed to the roof and squeezed into the available space. Alanna took the seat in front and I walked around to the driver's door.

Once we hit the road, driving from Berkeley, California, through the Sierra foothills to Wells, Nevada, we began to talk. The first leg of the journey to Jackson, Wyoming, took eight hours, and we told story after story. Ronni and I mesmerized Alanna with tales of our working as stewardesses between the Orient and Europe. She told us about growing up in Chicago.

When we woke the next day, the sun had announced itself in a bright blue sky. We continued on our way to Jackson, just south of Grand Teton National Park. I had contracted a booth at the annual fly-fishing event and would be marketing my calendars to individual attendees and fly-fishing corporations. On the first day of the show, a good-looking man approached my booth. I immediately noticed his muscular build, blond hair, and blue eyes. He was a tall, striking figure in his Western-cut corduroy jacket and cowboy boots. As president of the Rocky Mountain Council for Fly Fishers International, Mike Wolverton wanted to buy calendars for fundraising favors. While showing him the latest edition of the Angler's Calendar, we talked about our love of the outdoors, especially camping and fly fishing. I told him of photographing and backpacking in New Zealand, Africa, and the British Isles. He stared at me in surprise and simmering heat flushed my face. I turned from his penetrating eyes, wondering why I had reacted like that.

"Why don't you and the girls join me fishing tomorrow? I'll be with Ernie Schwiebert. You'll get some great photos."

"We have the morning off. Sounds like fun, but I'll check with them first." My co-workers agreed to join me while I photographed the two fly fishers on a trout stream beneath the majestic Teton Mountains. Two days later, we returned to California. Another successful selling event was now behind us. We began packing calendar orders and shipping boxes the very next day.

Chapter Two

Introduction to Sky Ranch

"Where're we going?" I asked.

"I want to show you the ranch," Mike said over the phone.

When I moved to Idaho the year after the Wyoming convention, my friendship with Mike Wolverton continued. I first managed Henry's Lake Lodge, near Yellowstone National Park, and then settled in Twin Falls and began working for the *Times News,* a local newspaper. Before long we started dating. On one of our first get-togethers, Mike asked me to meet him at the Murtaugh Cafe on Highway 30.

As I drove to the building, I saw Mike leaning against the passenger door of a blue pickup, his tight jeans embracing long legs, a rodeo belt buckle glistening in the sun. His boots, an off-white snap shirt, and a beige cowboy hat completed the look. He had an athlete's build, physically impressive from lifting heavy fertilizer bags and bales of hay. I parked beside his vehicle and walked to face him.

"It's a beautiful day. You'll enjoy the ride," Mike commented as he circled his arm around my shoulder.

He steered me toward his pickup and looked at the sky. Cumulus clouds meandered high beneath the blazing sun and cast moving shadows across the landscape. Mike drew deeply on a cigarette, its tip glowing before he flicked the butt to the ground. He crushed the cigarette with the toe of his boot and then opened the truck's door. I had never been inside a pickup. My high school and college friends only had cars, station wagons, or VW buses. No one had pickups, or trucks of any kind. This was another "first" for me.

Sitting next to Mike on the drive to Sky Ranch, I felt a warmth wash over me. How special he was as we traveled into a new setting—one that was so different from anything I had ever experienced. Mike pointed to a vast valley, a flat expanse of land rising to meet the massive sky. Hence, the ranch's name. He explained that besides owning a few hundred cattle, which roamed the corners of three states (Idaho, Nevada, and Utah), the ranch consisted of industrial, row-crop farming. The Wolverton family farm raised potatoes, its number one cash crop, along with wheat, barley, peas, beans, and alfalfa.

On our way east along Highway 30, I remarked about the difference between these farm lands and the tree-laden countryside of Connecticut. We passed cultivated fields, desolate prairies, and several farmhouses, but almost no trees, except those planted near houses or as pasture windbreaks. Tall silos, stationed near weathered barns, reminded me of New England lighthouses. Before long, I saw snow-covered mountains in the distance, far to the north.

"What's that?" I asked as I pointed my finger toward the horizon.

"Those are the Sawtooth Mountains. About eighty miles from here."

"Wow! We can see eighty miles," I exclaimed. "The air is so clear, no smog or haze."

"That's until harvest. Then the sky is covered with dust."

"Look at that," I exclaimed as I pointed to a nearby house. "The lawn is covered with water. Their pipes must have broken."

"No," Mike said. "They're flood irrigating. They opened a ditch and diverted the water over their lawn. Once it's flooded, they'll close the ditch. Very few places around here have lawn sprinklers."

"So, they flood their lawns and then wait a week or so to flood them again?"

"Yup. It's pretty easy to do and is much less expensive than sprinklers."

About fifteen miles past Murtaugh, we stopped at the ranch headquarters, a large, vinyl-sided building at the junction of four dirt roads. Inside, he showed me three maintenance bays, each large enough to handle the enormous grain and bean harvesters, known as combines. He pointed to one bay the ranch used specifically for painting vehicles. His employees accomplished this never-ending chore during the slow months after harvest and before spring planting. From there we sauntered into a rectangular office, a modest room with two grease-stained, swivel chairs and two small oak desks in the middle of the room, six feet apart, each with a black telephone resting on top. A cord curled from a rotary-dial phone to a plug in the wall. In the 1980s, only landlines existed. The office looked relatively sparse as Mike and his brother, Don, handled most farm business from their individual homes.

A storeroom on the side of the maintenance area amassed an assortment of parts. Row after row of cubby holes filled the walls. Each cavity held a variety of nails, screws, bolts, wires, tubing, and pipes in all sizes and shapes. On a pegboard hung countless tools and electrical instruments. Because the ranch existed so far from town, it had to be as self-sufficient as possible. Employees needed to repair items quickly and not spend three hours driving to and from Twin Falls, dropping off and then waiting to collect restored machinery.

Across from the ranch headquarters, a few single-family dwellings crowded together. These were the homes for several workmen and their families. Mowed lawns bordered the dirt road, and sagging clotheslines, laden with scrubbed garments, billowed in the breeze. From there Mike drove around various fields, explaining crops and irrigation systems to me. He stopped at a field, knelt in the earth, and sifted the soft soil between his fingers. I leaned on the truck's door and looked out the open window, watching him grope the brown dirt.

"What're you doing?" I asked.

"Just checking for clay and silt content," he answered. "The land of Southern Idaho is made of sandy soil. That allows water to wet our crops and yet drains good enough for air to circulate. That's why Idaho has such great potatoes."

As we maneuvered another mile down the dirt road, he pointed to a platform with several silver cylinders sticking up into the air. The center pipe took water from a deep well while the other cylinders leaned inward to steady the main structure. We drove from the road to the center, bouncing over deep ruts made by wheels connected to structural arms of the rotating center pivot.

"Come look at this," Mike said as he stepped from his pickup. I opened the door and followed him to the sturdy machinery. One arm protruded from the main water source. It sprayed water from various trusses moving on rubber wheels, circling through the newly planted field. I sat on the wooden platform, my elbows on my knees, and looked at Mike. He continued reciting facts about the different irrigating systems.

"Some wells drop about two thousand feet," he said. "But we only have to raise the water five or six hundred feet."

"Wow! I can't believe how deep it is," I exclaimed.

"This part of Idaho is basically a semi-arid desert. But we have a large aquafer below us. Using a well is much more efficient than irrigating with open ditches or siphon tubes," he said.

"What's a siphon tube?"

"It's an 'S' shaped tube," he said. "You stick the tube in a canal, holding the top end shut, and then drop it quickly into a crop's row. When a farmhand gets the hang of it, he can move water practically as fast as he can walk. I'll show you how it works on our way home."

"How many center pivots do you have?"

"Right now, we have twenty-two. We've been changing our fields as fast as we can afford it," he said. As he reached into his breast pocket and pulled out a cigarette package, shaking one loose and putting it between his teeth, he added, "Each pivot covers over a hundred acres."

Just then, the pivot's irrigating machine kicked on. A loud blast jolted the air and the platform began to shake. I must have thought I was going to plunge into the deep well. My legs rotated and raced as if I were on a bicycle. But I didn't budge. I couldn't get my butt off the platform. Mike came over and pulled me up, laughing so hard he spit out his cigarette.

"Where do you think you're going?" he said, hooting with laughter.

"That's not funny," I countered. "I thought I was going to die!"

"No, you're safe. Let's head to the gravel pit next," he said, snickering under his breath.

He ushered me back to the truck, and we traveled toward the middle of the ranch, turned north, and entered the gravel pit. Mike drove down a natural ramp formed of crushed rocks, leading into a large area filled with stones of all sizes.

"This is where we get gravel to shore up our roads. Whether it's for plowing or leveling roads, we take care of ourselves. No government agency working here."

Mike walked around the pickup and opened my door, holding out his hand to help me out. We strolled deep into the pit, surrounded by large boulders, and looked at the blue sky. He pulled me toward him, enveloping me with a kiss, as I wrapped my arms around his neck. It felt wonderful, like sunshine on water. After a few minutes, Mike pulled my hands from his neck and steered me back to the pickup. He opened the door and I glided over the vinyl seat. He slid in beside me. We were like teenagers . . . all moans, kisses, and body contact.

On our way back to the coffee shop in Murtaugh, he asked, "So, how'd you like Sky Ranch?"

"The ranch was fine, but the rancher was even better," I said as I snuggled closer. "I can't wait to see you again. Why not come for dinner? How does Friday sound?"

"Okay. I'll be there after work, about six."

I left him at the coffee shop and drove home, singing to myself along the way, "Six foot two, eyes of blue, but oh what those six feet could do. Has anybody seen my guy?"

* * *

By the time Friday rolled around, I had decided to prepare one of my favorite meals: spaghetti with homemade sauce. I had rented a two-bedroom farmhouse a few miles east of Hansen, about ten miles from Twin Falls. Mike would soon arrive for dinner and I wanted to impress him. After opening a bottle of Louis Martini cabernet to let it breathe, I placed wine glasses, silverware, and china on my dining room table. Setting a beautiful table would lessen any possible comments about my less-than-stellar cooking skills. I had not been raised in a culinary household. My mother only tolerated cooking; she considered it a daily chore. We had fish on Fridays, salads and casseroles on the other days, and usually went out for dinner on Saturdays. Our breakfasts were orange juice and cereal.

It was late in the day when I started preparing the sauce. I hurried from cabinet to cabinet, whirling around the kitchen trying to get everything ready for my special dinner. It was the first meal I would be cooking for Mike, and I was nervous. While the ground beef browned in a deep, frying pan, I rooted through the refrigerator and found a few onions to chop and threw them into the pan. Mouth-watering smells filled the kitchen. But when I checked the storage cabinet, I saw that I had no tomato paste. As Mike would be arriving shortly, there was no time to drive to town and shop. I improvised with ketchup.

That looks pretty good, I thought as I gazed into the pan. I took a long wooden spoon and stirred the red tomato sauce around the meat and spices.

Then I realized I had forgotten to buy tomatoes. Working at the *Times News* and owning an international calendar business meant my household chores were often left to the last minute. Or in this case, not at all. Without tomatoes, I improvised again. Outside my back door was an apricot tree. Being about the same color, I thought he wouldn't notice they weren't tomatoes. I picked a dozen apricots, washed, peeled, pitted, and cut them into bite-size pieces. Once they were added to the meat mixture, the apricots blended into the sauce. "He'll never know," I assumed. But just to be sure, I increased the amount of red wine to my strange concoction.

Hearing Mike pull into my driveway, I added grated Parmesan cheese to a crystal bowl and filled our wine glasses. He parked his pickup and walked to the backdoor, flicking a still-smoldering cigarette to the gravel driveway. Without knocking, he stepped into the kitchen.

"Smells great. What's for dinner?" Mike asked as he breathed in the onion and garlic and came directly toward me. He wore jeans, his rodeo belt buckle, a snap shirt, and cowboy boots. After a lengthy kiss, I directed him to the dining table and we raised our wine glasses to toast my first-ever dinner, cooked just for him. He sat with his long legs folded beneath his chair and told me about his day at the ranch and his rush to get here from his temporary home in Kimberly. When he started to eat the garlic bread, I understood how hungry he was. I rose from the table and brought in two plates filled with spaghetti and covered with my special sauce.

Mike smiled at me when I placed a dinner plate in front of him. Then he stopped smiling as he looked at the contents. Not sure what to do, he began to push the sauce from one side of the plate to the other. Finally, he bit into the spaghetti.

"What's this?" he asked as he poked at my imitation tomato.

"What do you mean?" I said as I twirled pasta onto my fork. "It's my specialty. It's a spaghetti dinner."

"What's in it?" he asked.

As soon as I mentioned I had replaced the tomatoes with apricots, he started to frown. He picked up another piece of garlic bread and pushed the sauce to the side of his plate.

"I'm sorry, honey, but I can't eat this," he said.

Deflated by his comment, I kept my feelings in check and told him we had ice cream for dessert. Knowing about my culinary inadequacy, it's a wonder he continued to date me. If he hadn't been so stubborn, he would have realized the special sauce was actually quite good . . . a little tangy but complimented perfectly by the savory, red wine.

Chapter Three

Cattle Ranching

"Want to join me while I check our cattle?" Mike asked a few weeks later.

"Of course," I said, sliding into the middle of his pickup's bench seat. Mike swung his right arm around my shoulders, and we began our drive through the foothills of the Albion mountains.

The cattle were located deep in Cassia County and spanned the corners of Idaho, Nevada, and Utah. He went often to oversee the operation, but I only accompanied him a few times. The price of beef had steadily dropped during the previous few years, and the family decided not to continue supporting the losing enterprise. They sold their remaining cattle the following autumn.

As we drove high into the mountains, Mike told me about his family. His relatives had been farmers and ranchers for generations. His parents relocated to Idaho after years of fruit and vegetable farming in Southern California. He was only nineteen when he joined his two older brothers, Gary and Don, at Sky Ranch. Gary, a strong, barrel-chested man in his late forties, was the oldest of the three boys. He moved away from the ranch and bought a machinery franchise in Twin Falls. He had an uncanny ability to fix all things mechanical from antique tractors to WWII airplanes. Don, the middle son, joined Mike and their parents, Margaret and Merle, handling and supervising the ranch responsibilities. Don and his wife, Georgina, and their daughters, Gina Dawn and Sarah, lived in a one-story brick home, a mile east of the parents' place.

Besides the three Wolverton families, Sky Ranch supported four other men, their wives, and their children. Terry Sherrill was the foreman, a mechanical engineer, and an overall whiz at repairing machinery. Melvin Tipton oversaw the center pivots. Mario Martinez and his cousin, Danny, were their backups. The four men maintained the many irrigating apparatuses, painted machinery, and kept miles of dirt roads relatively smooth and free from snow and floods. Mike and Don joined them during planting and harvesting and hired contract workers to weed and help with extra chores during harvest.

Mike maneuvered one of the ranch's pale-yellow pickups toward the high pastures. He had tied a red scarf around his neck and wore an off-white cowboy hat. We passed several gates on our way up the mountains. Mike stepped out to open and shut them while I slid to the driver's seat and drove the pickup through the wide opening. Soon we came to a cattle guard. I had never seen one and realized the object worked just like a fence. A vehicle could drive over the guard, saving the need for Mike to get out and unfasten a gate. The rancher had a depression dug in the dirt between two fence lines. Next a dozen metal bars or tubes were laid inside the recess, perpendicular to the road. He placed them about six inches apart and painted them bright yellow or orange. If an animal tried to cross the barrier, its feet became stuck in the contraption. On the whole, the cattle knew not to attempt a crossing. It worked just as well as a gate, keeping stock wandering from field to field.

At one span we came upon a Hereford bull whose foot had lodged between the metal tubes. The brown and white bovine bellowed in distress. Slobbery grass hung from his mouth as he tugged his wedged foot. There was no way Mike could get him out by himself, and his handheld Motorola radio did not work that far from the ranch. We crossed the cattle guard just inches from the angry bull.

"One of the cowboys will be checking the pasture today," he said. "He'll figure out what to do." And he was right. When we returned a few hours later, the bull was gone.

The trail narrowed and shrubs brushed against the sides of the truck. At a watering trough, high in the sage-scented mountains, a group of cattle had gathered. Most were black Angus, a breed derived from Scotland, but a couple were creamy white. The Charolais, a breed originating in France, mingled with the herd. As we sat in his truck surveying the cattle, Mike told me when Charolais are bred with Angus, the offspring are not as feisty and are easier to handle.

"Angus mothers are tough protectors," he said.

"What'd you mean?"

"I had one jump into the back of a pickup when we were doctoring her calf. Talk about panic time! When she jumped in, we instantly flew over the side rails. No way were we getting in her way!"

"What do you mean by *doctoring*?" I asked, smiling as I imagined the mother cow and two or three cowboys jumping in and out of the back of a pickup.

"Sometimes we'd rope a calf. And sometimes we'd put it in a squeeze chute. And if there's one small calf, we'd pick it up and swing it into the back of a truck."

"Then what?" I asked, so curious about all I saw. His was a lifestyle I knew nothing about, and I was interested in Mike's answers.

"We'd take its temperature and give it a round of vaccinations."

"What about the ear tags?" I asked, pointing to the plastic, yellow tags hanging from each cow's ear.

"That's a recording device. We use the tags instead of branding. See the number on each tag?" he asked.

"The one closest to us is Number 116," I replied.

"That's an identification number. It tells the animal's history, its lineage, past owners, and latest inoculations. It can also help track contagious diseases," he explained.

"Sometimes we'd accidently round up another rancher's cow," he continued. "With the number, we can get it back to its rightful owner."

Mike went on to say, "Besides doctoring the calf and attaching ear tags, we'd dehorn and castrate him."

"Wow! No wonder the animals run from cowboys," I said as I surveyed the herd, nodding my head as I heard his answers. "So, tell me, what's a squeeze chute?"

"Am I your own personal cowboy encyclopedia?" he teased before he continued to answer my questions. "See that green metal object to the right? It's a portable chute. We trailered it here a few weeks ago."

"Yeah. Then what?"

"It's adjustable and can control the calf by squeezing its sides. And it locks the head still. With the calf inside, we can do everything without hurting ourselves or having the calf break loose," he added.

"How do you castrate a bull?" I asked.

"A cowboy slices between the balls and uses a clamp to actually remove the testacles. He then places a special bandage over the gap. It not only closes the opening but also medicates the site."

"Gad, that makes me sick," I said.

"Not as sick as the bull feels," Mike laughed. "Haven't you heard of 'Rocky Mountain Oysters'? That's something Don likes."

"Oh, no. He doesn't eat them, does he?"

"Yup. He and some friends have a few drinks, roll them in flour, and deep fry them. He says they're pretty good."

"I'd need more than a few drinks to try that," I countered. "Nope," I amended my thought. "I couldn't even do it with liquor."

Confined in the Ford truck for a few hours gave us plenty of time to share life stories. I knew he had two children from a previous marriage and that they lived with his former wife. I told him of my previous ten-year marriage and numerous funny stories of fishing around the world.

Further down from the mountain as we bumped along the dirt and gravel road, Mike told me about the time a few years earlier when he and Lou, a hired cowboy, had trucked their horses to handle some doctoring high in the mountains. They wore leather chaps over their jeans and cotton bandanas around their necks. The chaps protected their legs from thorn and mesquite bushes. They used the bandanas to wipe sweat from their hands and faces and to cover their mouths and noses if an unexpected dust storm materialized. Surprisingly, they didn't use them to blow their noses. Instead, they would close one nostril and blow hard, aiming the nasal material into nearby bushes. For me, that's when the romantic image of a cowboy crossed the line. My father would have been disgusted by the practice. He always used a monogramed, Brooks Brothers handkerchief to blow his nose. Nothing less would do.

Mike continued to reminisce about his life. One time he and Lou trailered their horses but didn't bring a squeeze chute as there were only a few calves to catch. Throughout the day, Lou continued to rope calves' heads while Mike roped their hind feet. After each man had connected to the calf, they looped their rope several times around the saddle horn. As I learned, the men would "dally" or wrap the rope. Then their horses would back up, stretching the calf. This ensured a solid anchor as the calf had to pull against both the horse and rider.

"I'd jump off Shamrock and race to the calf, push it on its side, and wrap a short rope around its legs," he said. Once the calf was secured, they'd proceed with doctoring.

Mike told me about riding Shamrock, his quarter horse, one particular day with his border collie stalking the strays. All day long, Lou and Mike and their horses roped and worked the calves like well-oiled machines. It was toward dusk when the unexpected happened.

Lou lassoed the last calf around its head, and it stormed out in the opposite direction. Mike kicked Shamrock and off they charged to rope the calf's legs. Just then he heard a shriek. Mike turned back to see Lou holding up his hand. His index finger was missing.

"What happened?"

"My finger got caught in the dally. When the damn calf bolted, the rope burned off my finger."

"Christ! Get off your horse. I'll trailer them back," Mike exclaimed. "You get to the hospital."

"No. I'm not going to any damn hospital."

And he didn't. After loading their horses and returning to the valley, Lou went home and drank the night away. Mike checked on him later that night and again the next morning. He guessed the finger had seared shut from the rope burn and no infection had set in. Surprisingly, Lou offered to cowboy for Sky Ranch the following week, roping and doctoring as usual.

"When you move cattle from your summer grazing range, where do you take them?" I asked Mike as we continued down the mountain road.

"I'd hire a cowboy or two and round them up. We'd herd them into a cattle truck and transport them to a feed lot."

Along the Rocky Mountain highways, I passed many silver cattle trucks, high metal eighteen-wheelers with large air holes spaced every foot or so. And when I drove from Twin Falls to the ranch, I maneuvered around the Hansen feedlot. The lot, divided into several enclosures, held only dirt and manure with nothing growing inside. Workers threw hay and a specialized growth feed into a trough alongside the edge of the pens. Tractors pushed dark soil and manure into large mounds around the barren ground. The piles looked at least four-feet high with one cow invariably standing on top. "King of the castle," I thought. While in the feedlot, the beef cattle gained weight until they left for slaughter, adding to the operator's profit.

I continually increased my knowledge about ranching—either by listening to Mike's explanations or by being observant as I drove around nearby communities. Once I inquired about a "female heifer." Little did I know the word *heifer* is the name for a young female cow. After he laughed at my mistake, Mike told me about the birthing process.

"I'd wear a plastic glove that reached to my shoulder. It protected my shirt and gave me easy access to the womb," he said. "First, I'd position the pregnant cow into a head gate and put on the sterile glove. With the glove on, my hand could easily slip into the birth canal." Once inside, Mike knew if the cow needed help completing the birth. If so, he'd wrap a small metal chain around both front hooves and slowly pull, inching the calf outward. Once its hips cleared the cow, the calf would drop headfirst into the straw, covered with moisture.

"I'd clean out the calf's nose and mouth and then dry its body with a burlap sack."

Mike continued, "When the calf raised its head, the mother took over, licking mucus from the calf's body and encouraging it to stand. Once it stood and began nursing, I knew everything was okay."

Mike told me that beef ranchers describe their cattle as 'mother cows,' which meant the ranch had a cow and a calf. Instead of telling people how many cows Sky Ranch had, he explained, "I'd just say, a few hundred MCs. Ranchers understand."

I absorbed cattle information like a sponge, listening to Mike's stories, questioning him about specific items, and asking for more details. Since I was constantly curious and our relationship was still new, Mike happily obliged my interrogations. How different life on the ranch was from my youth. I had grown up in a neighborhood full of children, had animals as cuddly house pets, and swam beside sandy beaches along Long Island Sound. Nowhere had I ever experienced the broad expanse of prairie lands enclosing massive farmlands, devoid of any real population. I found the prospect of adjusting to this barren landscape appealing, even imagining my life with Mike as another great adventure. Little did I know what the future held for me.

* * *

On my first solo trip to Sky Ranch, I marveled at the quiet scenery. Individual farms were neatly organized in a mosaic of parcels, from orange to yellow to green and brown, like a muted kaleidoscope. Their colors scattered in the emptiness of huge fields, surrounded by a few huddled trees and bushes, planted not only as house decorations but as protection against the strong winds blowing through the valley.

After I drove across the Hansen railroad bridge, past the cutoff for Murtaugh on Highway 30, I looked for street signs. Nowhere did I see typical road designations, like Maple Drive or Smith Lane. Instead, I noticed only numbers. Mike had told me to turn right at 4900. Rural addresses in Idaho were county grid numbers, arranged by north, east, south, and west. Supposedly, if one knew the coordinates, finding the location was easy. In reality, almost everyone used visual directions. They'd say, "turn right at the green house" or "five miles past the old potato cellar."

When I saw 4900 on a street sign, I turned right and began to notice familiar houses and barns. There weren't many but I remembered them from previous trips to the ranch. Halfway down the street, the asphalt turned to dirt beneath my tires, a road that now resembled a waffle iron—with potholes and bumps along the way. I twisted my silver Subaru from right to left, dodging to avoid the worst of them.

While I continued on the gravel road, a large tractor came toward me. As we passed each other, I looked into the cab and saw a young boy at the wheel. He looked to be about twelve, yet he drove a huge expensive vehicle, probably transporting it from one field to another. It amazed me how farmers allowed, and even encouraged, their young boys to drive pickups and enormous farm equipment. I believed children should be more mature before permitting them to handle such huge vehicles.

On my return from visiting Mike at the ranch, I ventured into Murtaugh, a tree-shaded farming town a mile north of the highway. There were no traffic lights, no banks, and only one tiny gas station. A few commercial businesses comprised downtown: a pool hall, a small grocery store, a café, a post office, and a hardware store. The Community Building Supply store on Boyd Street reminded me of a tiny Wal-Mart. It had appliances, gifts, fishing and hunting supplies, farm goods, and hardware . . . all packed into one building. And it saved nearby residents from driving to Twin Falls if they only wanted to buy a few items.

From there I followed the old highway as it curved along the Snake River canyon. Tumbleweeds blew in front of my car and became snared in strings of barbed wire fences bordering the road. They stacked in bunches, so thick they formed a solid barrier. This was a typical sight in the blustery conditions throughout the prairie lands. The old road eventually merged into Highway 30 with the sun dipping from the sky as its last rays reached across the highway. I continued to drive west and pulled into my gravel driveway as dark dusk enveloped the land.

Chapter Four

Fly Fishing Honeymoon

On the afternoon of August 7, I became Mrs. Michael M. Wolverton. My parents flew in from Florida and joined Mike's family and friends for an intimate ceremony at the Methodist church, a small building tucked into a Murtaugh side street. I wore a light pink chiffon dress, fitted at the waist and falling to my knees. Mike wore a navy suit with a white shirt. This was the first time I had seen him wear something other than jeans. He wore his extravagant, ostrich-leather Tony Lama boots, now gleaming with dark brown polish.

Afterward we had a small celebration at our newly built home at Sky Ranch. Mike and I had designed the home from a myriad of house plans, and our furniture had arrived a few days prior. The decoration was sparse but our couches, beds, and chairs were in their appropriate locations. Mike carried me over the garage-door threshold and into the kitchen. The wedding party soon followed. Typical of many houses across Idaho, family and friends used the back door. Strangers and formal company came in through the front door.

The outside of our brick house was patterned after a Tudor home on Harrison Boulevard in Boise. It had a steep, cedar-shake roof, a protruding gabled dormer, and multi-paned windows. I paused and looked at the kitchen. It had a full refrigerator with a matching freestanding freezer, two ovens, a microwave, and a gas range. What a dream kitchen—amazing when one considered what a dreadful cook I was. Mike must have thought these appliances would inspire and contribute to my future culinary successes. I hoped he was right.

A black, handheld Motorola radio rested in an electric base on the white, manmade Corian countertop. It stood a bulky eight inches high and had a five-inch antenna extending from its top. I had never used a portable radio and had never even seen one before coming to Idaho. It wouldn't be long before I'd know the Motorola system by heart, but at the moment it looked formidable. The radio announced discussions from three other ranches besides our own, and I could hear their conversations as they could hear ours.

The stairs to the second floor started just to the side of the front door. Underneath the staircase a deep closet contained a garden hose attached to a faucet. Because any emergency vehicle would take almost an hour to reach us, we had to rely on our own water supply to douse a house fire. The living room had glass-fronted, mahogany cabinets on each side of a wood-burning fireplace. On the south wall large windows had unobstructed views of nearby fields, with

the opposite space opening to a step-down room dominated by a pool table and bordered by a wooden railing. Down the hall was Mike's office and a guest bathroom. The master suite accommodated the north end of the house. Cheerful sunlight streamed in through its three tall windows facing the South Hills. Upstairs were two bedrooms for Mike's children, Christa and Blaine, along with a bathroom, and a large office over the garage for the Angler's business. His children lived with their mom but visited us whenever their school schedules allowed. Amazed by this beautiful home and my husband's welcoming family, I envisioned a life of comfort and class—similar to the assets I had growing up in Connecticut. I was ecstatic.

Gathered in the dining room, family and friends held glasses of sparkling wine and listened to congratulatory speeches. Mike mentioned our next day's agenda.

"What? You're going on your honeymoon with your parents?" Del Carraway, one of Mike's fishing buddies, asked. Mike and I, along with his parents, were driving to Montana the following morning. I had our tent trailer packed with camping supplies and he had his pickup loaded with angling gear.

"What could be better?" Mike responded with a smile as he hugged my shoulder and pulled me closer. "Fishing with Dad and having Mom cook the meals."

Merle and Margaret had a bus-like, recreational vehicle and pulled our fourteen-foot Mirrocraft boat. Mike towed our tent trailer behind his pickup. We drove through Yellowstone and set up camp near a large lake on the Flathead Indian Reservation, north of the national park in Western Montana. As only our two vehicles were present, we had absolute privacy and excellent fishing. Margaret, a tall, slender woman in her late sixties, prepared the meals while I placed glasses and plates on the outdoor table and cleaned up afterward. During the day I fished and photographed with Merle. He had health issues so I ran the boat. On the last day, Margaret made a special dinner and wanted us to eat together. However, when the fishing was hot, it was difficult to coordinate the anglers with a specific time to eat.

"Mike's too far away," Merle stated as he spied his son at the remote end of the lake. "He can't see us waving."

Mike fished from a float tube: a nylon-covered inner tube. He rested on a mesh seat in the middle of the tube and wore belted chest waders. He had swim fins attached over his wading boots, allowing him to maneuver the tube wherever he wanted to go. Because Mike fished from such a distance, it would have taken him over an hour to paddle back to his parent's motorhome.

"Okay," I said from the rear of our fishing boat. "Let's pick him up."

Sitting near the motor, I held the rudder and propelled the boat across the large lake. Once I circled Mike's tube and pulled the Mirrocraft close to him, I cut the engine and drifted to his side. He grabbed the gunnel and hung on. Once he settled next to the boat, I slowly glided back toward the campground.

"Go faster!" I heard him shout over the roar of the motor. The thirty-five-horsepower Evinrude outboard caught immediately. I pushed the gear lever forward, gunned the engine, and we flew across the water. Massive waves plastered Mike's back and practically swamped him in his float tube. He clung to the boat's gunnel, his knuckles white and powerful, trying not to be swept under the rush of water. As soon as I saw him struggling, I cut the engine. Mike wiped his face, repositioned his fishing rod, and stared at me in disbelief.

"What the hell are you doing?" he shouted.

"You said, 'Go faster.'"

"No, I said, 'No faster.' Let's take it a little easier this time," he said as politely as possible. We took off again, but at a much slower pace. Merle turned toward me, a slight smirk filling his face. I think he thought the soaking of his son was quite funny.

Chapter Five

Making Meals

"How do you make over-easy eggs?" I asked Margaret over the telephone. I only knew how to cook scrambled eggs. She laughed as she answered.

"Put some oil in a skillet and turn up the heat. Drop eggs in the oil and try not to break the yolk," Margaret instructed. "You want the whites to be firm and the yolk soft," she added. "Flip them over and cook them on the other side, but only for a few seconds. Then they're ready. Don't worry. Mike has patience—especially if he's hungry."

While making the first breakfast in our new home, I felt intimidated by the process. Mike expected a hearty, three-course meal. I was not an enthusiastic early-morning eater. My own breakfast consisted of one piece of buttered toast and a glass of orange juice. For Mike's breakfast, I made coffee, poured a glass of orange juice, and prepared a half grapefruit. The citrus scent drifted upward as I sliced each individual wedge and placed the turned-up grapefruit on a small plate. The second course was a bowl of cereal and a glass of milk. When it came to the third course of bacon and eggs, I had to call his mother.

She reminded me that we lived over four thousand feet in elevation and everything took a little longer to cook. I fried two eggs as she instructed and they turned out just fine. Bacon drained on paper towels and two pieces of toast popped from the toaster. Butter and jelly sat on the kitchen table along with a gallon of milk and an ever-present container of toothpicks. To the side of his plate stood a mug of coffee. I felt proud of my accomplishment. I had actually completed his breakfast.

The smell of bacon drifted to our bedroom, its airborne invitation beckoning Mike to the kitchen. He started right in with his meal as he had a busy day ahead. After his first course, he scraped his bowl of Raisin Bran and wiped a dribble of milk from his mustache. Then he began his bacon and eggs portion. I had heard farmers were hearty eaters and that morning I saw one in action. Mike could go through countless calories when lifting heavy seed bags, carting bulky machinery parts, and working large equipment. Because his breakfast consisted of the same items day after day, I adjusted. Before long, it became just another acceptable ranch routine. When he finished, Mike leaned down and kissed me.

"See you at lunch," he said.

"Okay. I'll have it ready at noon," I called back over my shoulder as I began to scrape the dishes and load the dishwasher. Although the sandwiches would be ready as stated, I never knew when he'd actually appear. Farmers don't conform to regular work hours. Mike came home whenever his chores allowed him an hour's break. While he worked, I decided to plant tulips and daffodils around the two entrances of our circular driveway.

Three employees arrived later in the day and began to install a split-rail fence across the lawn next to the dirt road. They used an auger, or fence-hole digger, attached to the back of a tractor to dig the post holes. I had read about people getting their clothes caught in the rotating device, whipping at such speeds, their legs and arms were actually ripped off. It was one of the most dangerous items at the farm and I kept my distance. Mike appeared and supervised the actual installation. He then had his men shovel wide holes and plant vibrant rosebushes, alternating their yellow and orange colors next to the fence.

"These are beautiful, Mike. I'm so happy. Our house is actually becoming a home."

* * *

During the first few days after moving to the ranch, I decided to introduce myself to our nearest neighbor. The Moss family owned a ranch bordering ours, a mile or so from our house. I turned east from our gravel driveway, drove over the dirt road to their white brick house, and pulled in beside a dusty pickup. I stepped from my car and walked up the path to their front door. Candyce Moss, a young blond about ten years old, greeted me. Once I introduced myself, she pointed to the field adjacent to their lawn.

"Mom's on the tractor," she stated.

Gravel crunched under my tires as I steered to the edge of the field and halted. From there I watched a tractor approach me. I exited my Subaru as the tractor began to turn. Then it stopped and paused. The cab door opened and a blond woman emerged and jumped from the machine. We met in the field and shook hands.

"Hi. I'm Bobbi Wolverton," I said. "I married Mike last month."

"So, I heard," she said. "I'm Marsha Moss."

What a beauty. I couldn't believe she wore a full face of makeup with foundation, mascara, and bright pink lipstick and painted nails. This was a woman who had just been driving a tractor, turning over dirt in a dusty field! As she said, she had been "tilling" the soil. Her tractor had steel tines that pulverized the dirt and sod, getting it ready for planting. We chatted for a few minutes, exchanging pleasantries, and talked about each other's family.

"I have two children, a boy and a girl," she said. "You met Candyce."

"Mike has Blaine and Christa but I don't have any. When I moved to Twin Falls, I found employment at the *Times News*. I also own a calendar company."

"Wow! Two jobs. And I thought I was busy," she said. "I rep for a small fashion company in Salt Lake. I drive there every few weeks when I'm not in the field helping Dean. He's my husband."

At the end of our conversation, I asked her to visit. We agreed on a date and she returned to her tractor while I drove home to prepare dinner. Mike expected three hearty meals a day when we were at the ranch. Thank goodness, my calendar company required we attend five business conferences a year. That was our time to splurge and dine at fancy restaurants—and to give me some respite from my daily cooking chores. Mike and I also had fishing and photographing jaunts all over the world: from Alaska to Argentina, from Canada to the Caribbean, and from Hawaii to New Zealand. With these trips, we could be away from Idaho for three weeks at a time. If I had had to live at the ranch day in and day out, I would have burned out. The constant travel kept me energetic and our love blossomed.

Chapter Six

Industrial Farming

E ager to learn about Sky Ranch, I discovered numerous facts about farming, not your average garden-variety farming, but huge, industrial farming—the business they perfected at the ranch. First, I noticed the rhythm of the seasons and realized when planting actually began.

It started in late fall when workers churned the ground after harvest. I watched the stubble in front of their large vehicles change to chocolate-brown chunks of earth flying out behind. Next, the workmen in their machines came back through the field and planted winter wheat. During winter months, deep snows encased the Rocky Mountains, irrigating the crop and keeping it from freezing. Finally, in spring the lingering snow started to evaporate. Tiny green sprouts of winter wheat began to penetrate the soil. That's when other crop plantings began in earnest.

Sandy soil, hot days, and cool nights made for perfect growing seasons in South Central Idaho, known specifically around Twin Falls as Magic Valley. During early spring, I zipped up my jacket and stepped outside, into the cold March air. I watched fields crawl with machines, moving up and down, keeping rows as straight as possible. Farmers judged each other by how even their rows were. The straighter the row, the more accurate the planting and harvesting. Straight rows meant more money. Workers left their machines in fields, ready to begin planting the following morning. Surprisingly, no one ever stole any farm equipment as keys were kept in the vehicles overnight. What was really on a farmer's mind was water, not loss of machinery. Magic Valley received only ten inches of rain a year and meteorologists considered it a semi-arid desert.

"Forget about oil," Mike once told me. "Water will be the world's most important commodity." And he was right. Today, countries, states, towns, and individuals fight over water rights. The more fresh water, the wealthier the region.

* * *

When Sky Ranch contracted a crop duster some time later in the season, I would stand on our front porch and watch a small yellow plane take off and land just a hundred feet before me, using the dirt road bordering our house as an air strip. The plane circled our pastures, being careful to rise above telephone and electric wires, and then dropped low to spray a fine mist of fertilizer or herbicide. It was like seeing a professional air show, and yet, there it was—free entertainment, right at my doorstep.

Whenever a field had been sowed too many times with one specific crop, Don and Mike would substitute it with alfalfa. Known as a "rotating" crop, alfalfa hay contained a rich assortment of protein and minerals that impregnated the ground, making it better for a subsequent crop. Sky Ranch generally had at least three cuttings of alfalfa. With many dairy farms moving into Magic Valley, they had consistent buyers. Once the professional hay mowers arrived at the ranch, the alfalfa would be cut and then raked into long rows. Then the hay would "stand" or dry in the windrows covering the fields.

Often, I'd see the two brothers outside their offices, frequently with Terry and an agriculture or fertilizer consultant. They stood in a casual circle with their hands in their front pockets, looking down and scratching the dirt in front of them with their boot tips, a farmer's way of thinking and discussing. This was a special ritual I'd see time and again . . . nothing like a corporate conference in a city office, but just as productive.

After lunch one day, Mike looked out our dining room window, checking the nearest field. He stood with his hands behind his back, his baseball cap pushed back on his head. I knew he wished for clear weather, no rain or wind, as we just had a cutting of alfalfa.

"Wow! Look at that, Bobbi," he said as he pointed to a faraway, dark object.

I walked over to Mike and stood beside him, staring out the window. There in the distance was a brown elk traversing the hay field behind our house.

"Must be a young male, looking for a mate," he said. "Obviously, it's lost."

Mike turned and moved to the kitchen. He reached for the Motorola and alerted Terry.

"Hey, we have an elk, west of the office," he said. "Why don't you herd him toward the foothills so he doesn't get into trouble? Clear."

"Okay. Will do. Clear," Terry responded.

That wasn't the first time we had wild animals crossing our farmland. Coyotes frequently trotted over the fields, and one time we had a cougar, also known as a mountain lion, cross behind our house. The tawny beast looked to be about two hundred pounds. We saw him retreat from the ranch, getting smaller and smaller until he simply disappeared into the foothills.

Chapter Seven

The Hound Pound

A t the *Times News*, I relished in the energy of the work place, the frantic pace to meet deadlines. My job was to obtain display advertisements from local businesses, create ads, and deliver them to the paper's design or typesetting department. Carol, my assistant, proofed the rough ads with the appropriate customers and brought them back to our office for me to either change or hand into the final production team.

She lived by herself and often talked about owning a small dog as a companion. It had to be able to ride in her car when she ran newspaper errands or to stay at home for hours on end. And it couldn't shed.

"Why don't you go to the pound?" I asked. "The building's right down the street."

"Will you come with me? We can go after work," she said.

"Sure. I've never been to one before."

Dressed in office attire, we drove together to the gray, cinder block building. The facility was dirty and dingy with thick weeds circling the place. Entering, we noticed the strong odor of animal waste. It completely filled our nostrils, and we practically gagged in response. Carol told the manager, a dark-haired man in his mid-fifties, what type of dog she wanted. He waved his hand, indicating he wanted us to follow him into the adjoining room. Our ears were immediately assaulted by profuse barking and uncontrollable howling.

As we walked the cement center aisle, we looked at both sides of the chain-link, fenced kennels. The dogs wagged their tails against the wire confines, barking to be liberated from their cramped prisons. Carol found a small mutt who jumped up and seemed to smile at her. She knelt in front of the cage, placing her fingers to the metal enclosure. The black and white dog looked like a terrier and poodle mix. When the manager opened the kennel door, the young dog leaped into Carol's arms and began licking her face. It was love at first sight. After cuddling the squirming pooch for a few minutes, she replaced him in the kennel and shut the door. As we walked away, we heard a howl and saw it jumping at the enclosure as if to say, "Don't leave me!"

Standing in front of the office counter, Carol filled out the appropriate paperwork and asked, "How much do I owe?"

"Fifty dollars," the manager responded, wearing a dirty, white shirt splashed with grease. "And it has to be cash. No checks or credit cards."

Carol and I stared at each other in complete surprise. This seemed to be a huge amount of money for a pound dog, over a hundred and fifty dollars in today's currency. And to demand cash only? We examined our purses but together we could not come up with fifty dollars.

"I'll bring cash tomorrow," Carol stated. "I can't get the money now. The banks are closed. It's after five."

We left the facility, feeling deflated. Carol had planned on bringing home a dog that day. She had already bought a collar and leash and had a month's worth of food.

During work the following morning, she chatted and laughed about her impending life: walking, feeding, and watching television with her four-legged companion. At lunchtime, Carol withdrew fifty dollars from her bank and placed the money in an envelope, marked,

"For Toby," the name she had chosen for her rescued puppy. She was so excited she couldn't keep her eyes off her wristwatch. Once the workday ended, she stood and shouted, "Let's go!" She was in her car, leading the way. I trailed behind in mine. We braced ourselves for the smell and noise, but not for the manager's announcement.

"You should've had the money yesterday," he said as small bubbles of spittle clung to his lips, his words raining down on us like a bad storm. "We can only keep animals forty-eight hours. Then I have to gas them."

"What?" Carol yelled. "I told you I'd be back. I have the money right here."

"People say that all the time and they never come back," he countered.

"Fine. But you could have taken a chance. One more day wouldn't have made a difference," she said as tears filled her eyes. "I can't believe you killed him."

I was shocked by the manager's complacent attitude. He had a routine of killing animals on a forty-eight-hour schedule and wasn't going to bend the rules. I was furious! Disheartened but determined to buy a dog, Carol walked into the back room. Although saddened by the loss of "Toby," she picked out another small dog, tan and brown with curly hair and a yipping bark. A new Toby had entered her life.

Tom Courtney, the Twin Falls city manager whose wife was a card-playing friend since first arriving in Idaho, received a telephone call from me the next morning. Still angry over yesterday's situation at the dog pound, I told Tom about the incident. I asked what the city could do to correct the situation.

"The city has no money. We can't hire any more employees," Tom responded. "If you want to do something at the shelter as a volunteer, go right ahead."

"Okay," I said. "Tell me what I need to do to be official."

"Come to my office. You'll have to sign some forms. Just keep me in the loop. I want to know your plans."

I arrived at City Hall during my lunch hour and we discussed several options. My parents were both active volunteers in Connecticut and I thought this would be a good way for me to not only rectify a wrong, but also to give back to the community. This would be my first time volunteering as an adult.

After visiting with the editor of the *Times News* about the various negative issues at the City Pound, I was encouraged to write an article about a volunteer program that could alleviate the situation. It appeared in the Sunday paper and thirty people signed up to support the cause. We met at the pound on Saturday, all in work clothes and ready to scrub the interior. We also picked a new name. Instead of the Twin Falls Pound, we changed it to the "Hound Pound."

The shelter manager was exceptionally pleased with our cleaning efforts.

"I can't keep up," he said as a way to offset the poor condition of the building. "I have to capture loose animals. And I have to gas all the animals exceeding forty-eight hours. Then I have to take the bodies to the rendering plant. I have no time to clean the place."

Hearing his woes, I telephoned Dr. Marty Becker, a personal friend and our family veterinarian. He currently works on the Dr. Oz Show as America's Veterinarian and is a national newspaper columnist, but in the 1980s, he was a small-animal vet in Twin Falls. He met me at the pound after work and we decided on a once-a-week schedule to euthanize animals, not every forty-eight hours.

"Gassing is cruel and inhumane," he said. "I'll get some other vets to perform intravenous injections."

"I know some, too. I'll ask them to help," I added.

Once Marty came on board, every veterinarian in Twin Falls joined the mission. We asked them to donate their time once every six weeks. It was an easy fix to an awful problem. In the meantime, the volunteers at the Hound Pound encouraged new pet owners to have their animals spayed or neutered and gave a handout with the

names and telephone numbers of local vet offices. For their volunteering efforts, the veterinarians gained many new customers. It was
a win-win situation, and we were all proud of our contributions. The
Times News covered the changes at the Hound Pound with a full-
page article. A few days later, the local Boy Scouts called and asked
if they could help.

"We'll buy the paint and ten boys will be there Saturday morning," the scout master told me. "They'll each receive a merit badge."

The scouts were at the building early on Saturday with brushes
and paint. They white-washed and painted the whole building,
inside and out. It looked amazing, fresh-looking and shiny. As they
gathered their supplies, a few of them started to pull weeds. Before
long the whole group of scouts circled the building and removed
every weed.

"Thank you so much," I exclaimed to the boys and their scout
master. "You not only did a super job, but you went overboard.
The building and grounds look fabulous. You definitely deserve your
merit badges."

Later in the week, I asked the *Times News* graphic artist to paint
the name, Hound Pound on the side of the building. That evening
he painted the pound's name and, as a surprise, brushed six-foot-
high dogs standing upright, marching out the front door as if in a
parade.

Volunteering after work one day a week, I soon discovered most
farmers and ranchers do not allow cats and dogs into their homes.
Their animals are not considered pets but working animals. Cats
catch rodents and dogs herd stock or retrieve hunted birds. After
much begging, Mike allowed our dogs into the house, but Buddy,
my black cat, had to stay outside. We had a pet door allowing him
to come into the garage for shelter and food. Sometimes when Mike
left for ranch duties, I'd sneak Buddy inside. We'd cuddle on our
loveseat while I'd read a book, with me stroking his back and rubbing behind his ears. As I read, I listened for the sound of the garage

door. The noise signaled that Mike would soon enter the house and I'd have to snatch Buddy from the couch and put him outside. I'd laugh as I looked out the window beside the front door as he shook his head as if wondering what had just happened.

Living in a male-dominated household was counter to the way I grew up. My parents were basically equal, with differing political opinions but compromising on day-to-day activities. At this time in Idaho and being in a farming household, I adjusted to most of Mike's wishes, but sometimes, I simply couldn't help myself. I was too independent. Pets had always been a part of my life. Consequently, I allowed Buddy inside whenever I knew I had a few hours to myself.

While working at the *Times News*, my boss assigned the Twin Falls Rendering Plant as one of my advertising accounts. It accepted dead dairy and farm animals as well as cats and dogs from the Hound Pound. The rendering firm processed these carcasses by separating the body from the bones and fat, and the fat into lard and grease. The meat would be ground into tiny pieces and sold as food for nearby mink, fish, and poultry farms. When I asked Mike about dead cows I'd see periodically along Highway 30, he told me farmers would haul them to a roadside ditch and notify the rendering plant. The company had its workers collect the dead animals and transport them back to Twin Falls for processing.

"Another fact about farming most city folks don't know," I said.

"Yup. The birth and death of animals are all part of a farmer's life," Mike said. And so it was. I remembered not to get attached to any animal that might someday become food. And to never name such an animal. How horrible would it be to say you were having Johnny or Betsy for dinner?

Chapter Eight

Florida Parents at Sky Ranch

"Any chance you two can fly back to Idaho for harvest?" I asked my parents over the phone.

"When's that?" Mom inquired.

"Any time from the middle of August until the end of October," I said. "Why not come in September?"

"I'm looking at our calendar right now," Mom said from their new Florida home. "How about the end of September?"

"Great. I'll let Mike know."

Having my parents join me during my first year at Sky Ranch was exciting. I'd be able to brag about all I'd learned and they'd be able to witness an industrial harvest. They wouldn't believe the size of the potatoes, the quantity produced, and the processing procedure. I couldn't wait to tell Mike.

By the time my parents arrived at the Twin Falls airport, Mike and I were waiting. We rescued their luggage from the tarmac and walked back through the terminal. Mom and Dad followed us from the building and mentioned their turbulent flight over the mountains. When they arrived at Mike's truck, my mother struggled to climb in. This was probably her first time in a pickup as it had been for me three years earlier.

Once settled in our home, I showed Mom how to position Mike's cowboy hat.

"You put it upside down. Then the brim doesn't bend," I said as I took Mike's hat and placed it on a shelf in our hall closet. She was impressed by my "cowboy" knowledge, so different from anything she had ever experienced.

Soon Mom took over our laundry chores, and Dad followed Mike while he completed his ranch duties. My father had spent his teenage summers on a small farm in Michigan and thoroughly enjoyed learning the specific intricacies of Sky Ranch's corporate operation. While Mom picked yellow and purple wildflowers along our roadside a few days later, I drove my father across the ranch to join Mike on a combine. When we arrived at the correct place, I saw Mike's combine traveling toward the end of the grain field with dust churning up behind the large, red machine. It would be at least an hour before he'd return to us, waiting on the side of the field.

"We'll catch him, Dad," I said as I drove my Subaru across the already cut grain, brittle with stalks. In spite of my lack of farming knowledge, I knew not to drive through the section that had yet to be cut. That was obvious. We bounced over center pivot wheel ruts, and my father hung onto the car handle above his side of the door.

"I don't think you should be doing this," Dad protested.

"Oh, it's okay. Mike will be glad to see us."

Mike stared straight ahead when I approached him on the left side of his combine. I waved, all smiles, thinking how surprised he'd be. And surprised, he definitely was. But not pleasantly. He reacted with alarm and quickly stopped the combine.

He glared at me. "What're you doing? You could have set the whole field on fire!" He would have said much more but my father stood beside me.

"I told her I didn't think she should be driving across the field," my Dad said in his own defense.

"I thought you'd be happy to see us," I whimpered. "I guess I didn't know the impact. I'm sorry."

"Okay. Come on, Jim. Climb aboard."

"Here's a snack." I handed over a lunch bag of sodas and cookies.

Mike leaned over and gave me a kiss on the cheek and took the insulated container.

"Just don't do it again," he whispered into my ear as he gave me the perfunctory kiss on my cheek. He straightened up and helped Dad climb the steep metal steps to the combine's cab. Obviously, I needed to know a lot more about farming. It's just a shame that I seemed to learn so many things by my mistakes.

Once I returned home, I realized my mother had already arranged wildflowers in vases around the house and had set the table for dinner. She placed silverware, napkins, and glasses for water and milk on the dining table.

"I just can't add the toothpick holder," she said. "I think it's disgusting the way people pick their teeth at the table."

"Oh, Mom. Not to worry. Most of the men only take them to use when they leave the meal," I countered. "I've yet to see anyone actually use them when they're eating."

"Okay. But it just gripes me," she said. "You'd think they were born in a barn—or in this case, raised on a farm."

We both laughed at her comment and retreated to the kitchen to prepare the rest of the dinner: steak, salad, bread, potatoes, and the ever-present pitcher of milk. For the next few days of their vacation, we settled into a schedule of Mom helping me at home and Dad following Mike around the ranch. A day or so later, we heard loud banging at the front door.

"What's going on?" I asked as I ran from the kitchen.

"Your potato cellar's on fire!" A strange man in his forties and wearing a cowboy hat pointed to our ranch's headquarters. I left the door open and raced back to the kitchen. As he followed me, I picked up the ranch radio. Mom left the laundry room to see who had arrived.

"Mike, the potato cellar's on fire! Clear!" I yelled into the radio. My announcement went out over the airwaves to our ranch and to three other farms.

"Okay, Bobbi. Calm down. Nothing's wrong," Mike said. "We just sprayed the cellar with chemicals. I'll explain when I get home. Not to worry. Clear."

"I'm sorry," the man behind me said. "There was so much smoke coming from the cellar, I assumed it was a fire."

"That's okay," I said. "But I'll certainly hear about this tonight. I'm sure the other ranchers are having fun after listening to another of my gaffes."

"Sorry about that. My name's Joe Tugaw. I run a few cattle south of here. Obviously, I don't know much about potatoes."

"Thanks, Joe. I'm Bobbi. Mike and I were married last summer. This is my mother, Florence Phelps. My parents are visiting from Florida. And as you can tell, I don't know much about potatoes either."

When Mike arrived home later in the evening, he told us Joe was a veterinarian from Utah and owned a small ranch in the nearby foothills. He added that Terry and Melvin were still laughing about my calling "Fire" over the radio.

"How was I to know? Joe was wearing a cowboy hat and seemed to know what he was talking about."

"That's okay. Just another lesson," Mike smiled and joined my mother and father at the dining table. The following day my parents returned to Ft. Myers, and Mike and I settled back into our daily habits.

* * *

For my first Thanksgiving dinner at Sky Ranch, Mike invited his family to join us at our new home. His brothers, their wives and children, and Georgina's parents and her brother's family would get together at our place. Christa and Blaine chose to have Thanksgiving with their mother. Then they could spend Christmas at the ranch with their grandparents and not hurt anyone's feelings.

Our moving boxes had been unpacked three months earlier and our home was decorated. While waiting for the big day, Mike and I placed six leaves into the dining table. He left to gather chairs while I set a formal table. With twelve gold-colored chargers under each dinner plate, I added china, sterling, crystal, candles, and flowers—but no toothpicks. The table looked exquisite and inviting. The children's card table was similar but without the candles.

Inside my large, New England lobster pot, I added ten potatoes, peeled and quartered, and put them on the stove at a low setting. In a pan I combined a stuffing mixture, not from scratch but made from a box. Cinnamon rolls were soon to be baked and sat in a corner covered with wax paper and a towel, rising for their second time. This holiday recipe came from my father's mother who was a wonderful baker. Her husband, my grandfather, had been a flour salesman in the early 1900s. On the other side of my family, my maternal grandmother barely cooked as she had staff to aid with kitchen chores. My mother's background explained her void of culinary skills and thus, my lack of learning from her.

I stuffed a twenty-four-pound turkey with chopped onions, oranges, and apples. I covered the bird with aluminum foil, and we drove to his parents' home, on the ranch three miles away. As if God were weaving a tapestry of colors right before our eyes, the hills had turned into reds, greens, and bright yellows. Numerous tumbleweeds blew across the road as we approached their house. It was another gorgeous, but blistery, autumn day.

Wearing our holiday finery, the women donned aprons over their dresses and helped Margaret in the kitchen. She had fixed her silver hair in tight curls and wore a flowered smock apron over her party dress. During the previous few days, she had made pies, cookies, and cakes from scratch: chocolate potato cake, chocolate chip cookies, apple and pumpkin pies. Sweet potatoes baked in her oven; a covered salad sat on the counter; peas, corn, and cranberry sauce were in separate containers; and a green bean casserole sat warming on her stove. At our house, we had the turkey, stuffing, potatoes, and cinnamon rolls. Plenty of food for twelve adults and six children.

"When are we eating?" Mike's father called from their game room in the basement. The men had been playing pool while the women worked in the kitchen.

"Okay. We'll leave now," Margaret answered as she acquiesced to her husband's suggestion.

"We'd better take everything," Margaret said as the day dwindled to late afternoon. "I've yet to make gravy." All the women took food containers and carried them on trays to their vehicles. A line of cars left my in-law's place and paraded to our house. They eased into our driveway, one after another. Mike and I reached our back-door before the rest of the clan. I walked into the kitchen while Mike greeted everyone as they entered, so proud to show off our new home. I strolled over to the top oven, opened the door, and raised the aluminum foil cover. There sat a beautiful turkey just as I had placed it two hours earlier: a creamy pink bird with no glossy brown skin. The oven was cool to the touch.

"But, Mike, I turned it on," I said as I twisted to face him. "Look, you can see. It's set for 350 degrees."

"Yup, you did. It's brand-new, so maybe it wasn't wired correctly," Mike said, sympathizing with me.

"Let me look," Margaret said as she walked toward the oven. The rest of the guests had unloaded their coats in our bedroom, and the children were laying out a number of games they had brought with them. The men knew where we stashed the liquor and began filling their glasses while we held a cooking conference in the kitchen.

"You're right, Bobbi," Margaret said. "You set the temperature, but you didn't turn the second dial to the 'bake' position. Didn't you notice the lack of smell?"

"No. We were at your house when it should have been cooking," I said, tears forming in my eyes.

"Don't worry," Georgina said as she hugged me. "Let's have some more wine and watch the guys play pool. Dinner will be ready before you know it."

By the time the turkey was finally cooked, the family was exuberant—laughing and joking at every subject. With all the liquor flowing, I think that Thanksgiving became one of the cheeriest and most energetic ever. We sat around the table and Gary led the blessing. Another harvest was over, the crops were sold at good prices, and the new year looked promising.

Chapter Nine

Duck Hunting

"Where'd you learn to shoot like that?" Mike asked after I shot a duck and watched it fall from the sky.

"My father took me to a hunting school when I was in junior high. It was put on by the NRA," I said. "That was just a lucky shot."

Mike was an excellent hunter, and I liked to join him on his quest for ducks. During our first hunt with Sam, a hundred-pound black Labrador I had rescued before marrying Mike, we drove to Bell Rapids, a section of the Snake River near Hagerman. While Mike and I donned our waders and loaded shotguns, Sam disappeared. We called and called for him. Nothing but the pounding of rushing water cascading over rocks could be heard. Then we saw him. He had crossed the rapids and was coming back toward us, swimming through solid white water, with only the top of his large square head and wide eyes showing. As soon as he made it to our side, he emerged and shook a thousand beads of water from his coat. I ran my fingers over his muscular neck and broad sides. I knew from then on that he was absolutely safe. We no longer had to worry about him while we hunted near those swift rapids.

Mike had given me a Browning over and under shotgun for Christmas the year before. It was slim and light but offered a decent kick when it fired. I knew to stuff the stock butt into my shoulder and brace for the recoil. My shooting skills were limited, but I enjoyed being outdoors and experiencing the hunt.

Within a few minutes we saw ducks fluttering in the distance. We hid behind some bushes, keeping Sam close to our bodies. Mike shot his limit of six and I shot three. But what was really amazing was when I had my best shot ever. I raised the barrel directly over my head, inhaled, let out some breath and squeezed the trigger, leading the duck by a few feet. It worked. The duck fell from the sky into a patch of cattails, and Sam retrieved it as soon as the bird hit the ground.

When we returned home, Mike plucked the ducks in our laundry-room sink while I chopped onions, oranges, pears, and apples and added them around the bird breasts inside my ceramic roaster. At 350 degrees for about an hour, they turned out perfectly. Because I basted game birds with fruit, the resulting dinner was exceptionally moist and tender, as juicy as a ripe tomato and as rich as a luscious steak. My cooking skills were slowly improving.

On another of our hunts with Sam, we drove to the "firing line," a section near the National Wildlife Reserve in Hagerman. Birds by the thousands rested in the Reserve on their way to and from Canada. At home, Sam went wild when he saw us getting out our guns. He raced around the house, in and out of Mike's office, and out the kitchen backdoor. Sam knew he'd be invited to go with us, and he loved to retrieve.

Once we pulled into a level spot near the Reserve, we began the ritual of putting on waders and getting our guns ready for shooting. We heard duck chatter above us and the whistle of wings as the waterfowl approached. And once again, Sam disappeared. In the fifteen minutes it took us to prepare for hunting, Sam had left and retrieved twelve ducks from the river shores. We had our limit and had not even taken a shot!

"Shit! Look at this," Mike grumbled as he surveyed the shoreline. He retrieved a plastic leaf bag from his pickup and snatched the dead ducks from the pond edge and placed them in a large bag. We then removed our waders, returned our guns to their cases, and climbed back into the truck with Sam right between us. He seemed to be smiling as he looked out the front window, maybe thinking what a wonderful job he had just accomplished. The pungent smell of wet dog permeated the pickup, and Sam was immediately relegated to the outside truck bed. Mike pulled a cigarette pack from his breast pocket and lit up. He backed out from our spot on the firing line and headed toward the ranch, muttering under his breath.

Mike buried Sam's ducks in the farm dump while I prepared dinner. Over our hamburger meal, we talked with disgust about some hunters and their senseless attitude toward wildlife. Because so many shooters tended to hunt without a retriever, dead ducks littered the edge of the Snake River. We felt sickened by the unattractive scene as well as the waste of ducks.

When observing wildfowl in the air, I noticed duck wings flutter super-fast while the Canada goose tends to maneuver in a slower motion, squawking as it flies in a V-shape pattern. I never hunted geese since they mate for life. When one is shot, the "spouse" flies down to try to find its mate. It then becomes a target and is often shot.

Throughout Magic Valley hundreds of geese inundated local corn fields, a prime feeding territory. I enjoyed several goose dinners at the Twin Falls home of my good friends, Nancy and Doug Strand. Nancy was the first person I had met when I moved from California to Idaho. Even though I later relocated to Sky Ranch, she and Doug continued to include us in their fishing, hunting, and social activities. She taught me to cook goose, by stuffing and surrounding the body with fruit and onions. Doug also showed me how to grill elk meat. Using his recipe, I roasted venison after sautéing the deer meat in garlic butter and red wine.

Mike only hunted deer the first few years of our marriage. Then he switched completely to bird hunting and fishing. We always had an abundance of food at the ranch. Not only did we have meat from a ranch cow, but we also had unlimited potatoes, beans, peas, and onions that I stored in our small root cellar underneath the basement stairs. The ranch crops were fresh from the ground and unbelievably delicious.

Chapter Ten

Ground Blizzard

"We're supposed to have a ground blizzard today," Mike cautioned as he stood on the kitchen steps leading to the garage. "Be careful."

I wondered what that meant as I backed out and headed to Twin Falls. I had now lived on the ranch five months and had never heard the expression "ground blizzard." I imaged a heavy snowstorm—similar to what I had experienced as a youngster in New England.

Once I had settled in at my desk at the *Times News*, snow started in wide flakes, drifting steadily as it fell outside my office windows. The world began to turn white and the newspaper manager let everyone off early, giving us the warning, "Get home before the worst of the storm hits."

Wearing a wool suit and a winter coat, I wrapped a plaid scarf securely about my neck and left the department. With my leather dress boots crunching on frozen snow, I hurried toward my car parked behind the building. A brutal wind stabbed ice crystals into my face; it felt like stinging nettles against my cheeks. I wiped my gloved hands over the windshield, removed most of the accumulated ice, and jumped into my trusty Subaru. It had a standard transmission and studded snow tires. I wasn't concerned. With guidance from OK Tires, one of my newspaper accounts, I had had four studded snow tires placed on my vehicle a month earlier. Driving down Second Avenue to Highway 30, my windshield wipers valiantly whisked snow to the side. I detected few cars still on the road. The city looked like a ghost town. Most people had heeded the county-wide warning and were already in their homes. On my car radio, the weather broadcaster stated, "Winds could top sixty miles an hour."

Away from the shelter of town buildings, snow slashed through the rays of house lights as my wipers struggled to keep up. When I stopped for a red light at Eastland, the wind intensified and the snow blew sideways. Trees bent from the strong gusts and my two-door sedan swayed with every blast. I managed to stay on the road, but not like several others. A pickup had overturned in a gully and a dozen cars had slid off both sides of the road.

As I continued to drive, I spotted a few vehicles coming toward me, their headlights flickering through curtains of the blowing blizzard, their lights weak from the accumulating snow. When I pushed past the town of Kimberly, I became the only driver on Highway 30. The view of the countryside had by then become obliterated by a thick blanket of swirling snow. I saw only thirty or so feet of road ahead of me and I reduced my speed.

An hour passed before I reached 4900. I shivered from fear, not from the cold. The heater in my car was remarkable and my toes felt toasty warm. But as soon as I made the turn south toward the ranch, I could no longer see anything, not even the road. I was in a complete whiteout. My hands tightened around the steering wheel. And I was scared. Scared to death.

The wind slammed the snow so hard and fast across the surrounding farm land, the only thing I spied were the tops of telephone poles bordering the road. I shifted down and put my car in second gear. Inching my way a few hundred feet south, I crouched over the steering wheel and looked straight up and to the right. I wanted to be sure I stayed about ten feet to the left of the dark, wooden poles lining 4900. Finally, I came to the side of a thirty-foot-high stack of hay on the west side of the road. For its entire length, there was a miraculous clearing. The haystacks had blocked the blowing snow. I could finally see the road as I stared over the dashboard. There in front of me I saw a stopped car. I pulled behind it and waited. It never moved.

After pausing several minutes, I pulled my collar up, protecting myself from the biting wind that whipped around the hay stacks and ran from my car, squinting in the storm. I knocked on the driver's window and stood freezing in the bitter cold. A young couple sat huddled inside. Its driver rolled down his car window, staring at me in disbelief. He told me he had radioed his family to let them know they were almost home. Being new to the area, I didn't recognize them.

"I live about four miles down the road," I shouted over the howling wind. "I'll follow you."

Once we moved away from the protection of the hay stack and into the blizzard, my car began to shake. Heavy gusts battered its side. I gripped the steering wheel and plowed ahead, the dirt road frozen into ruts as hard as granite. Once again, I could see nothing in front of me, not even the car I had hoped to follow. It had been swallowed up by the storm. I was on my own. It was like driving through a thick fog, so dense I felt a white carpet had been thrown over me. Thank goodness, the tops of the telephone poles could still be seen.

With a sigh of relief, I finally recognized the exit road to our part of the ranch. Before I made the turn, I stopped and shifted into first. No way did I want to slide into a roadside ditch. Once I headed east, the wind came from the back of my car and blew the Subaru forward. I shifted to second and then to third. Finally, I made it home and into our garage.

"Where have you been? I was worried," Mike said as he stepped from the kitchen into the garage. "And who was following you?"

"Following me? I have no idea. I couldn't see anything."

Just then the telephone rang. Mike picked up the receiver and heard that Kay and Bill Nebeker had ordered one of their workmen to follow me. They wanted to be sure I had made it home safely. I never noticed anyone behind me. The blinding snow had been too strong. The Nebekers must have been the young couple in the car that had stopped alongside the haystack. I told Mike of my adventure from Twin Falls and all the trucks and cars on the sides of the roads. I also mentioned, "I've never heard the words 'ground blizzard' before."

"Now you know," he said. "It's an unusually strong wind that picks up the ground snow and blows it sideways. Often it isn't even snowing. Just blowing."

"Because we have so many trees in New England, the snow never gets a chance to blow across the ground," I explained. "Storms are definitely different here."

"I'm glad you're home. Safe and sound," Mike replied as he helped me bring in the groceries. "So, what's for dinner?"

Because of the ground blizzard, most businesses had shut down early that day, as had all Magic Valley schools. Right after lunch, Murtaugh bus drivers transported the school children home. A driver left the two Moss children, our neighbors, in front of their residence and drove away. He didn't know Marsha and Dean had locked the house while they were shopping in Salt Lake City. The children had no keys, and there were no houses within walking distance—especially during a blizzard.

While the snow continued to lash across the farm land, Candyce realized she couldn't get inside her home. She convinced her brother to join her inside their dog house. The kids huddled with the wolf-huskie mix dog and stayed warm with only a little discomfort as they waited for their parents to come home. From then on, Marsha hid a key outside the house. In case of an emergency, her children could open a door to get inside.

Chapter Eleven

Pregnancy and Birth

"Your breasts seem bigger," Mike whispered as he caressed me under our bed covers. When he pulled me closer, I contemplated his statement.

"What the hell does that mean?" I thought.

I had been nauseated a few weeks earlier and had left the *Times News* promptly to drive back to the ranch. I thought I had the flu. Many employees had called in sick as the virus had enveloped Magic Valley, forcing several schools and businesses to close.

Years before, I had married a college sweetheart. We were open to having a child but after ten years of trying, we never did. Consequently, I thought I never could.

At noon, I walked from the *Times News* to Main Street and down three blocks, past the Bank and Trust building and Hudson Shoes to Crowley Pharmacy. I bought a pregnancy test. The result showed that I actually could be pregnant. But, the brochure stated, the results were only ninety percent accurate.

"I must be one of the ten percent. There's no way I'm pregnant. My gosh, I'm forty," I reasoned. Mike and I had discussed children before we married. He already had a son and a daughter. He concluded two children were enough.

"Not to worry," I told him. "I can't get pregnant."

After an appointment with my gynecologist, Dr. Miller, he said, "yes," I was indeed pregnant.

"How can that be?" I asked.

"It's the same as when a couple can't get pregnant, and consequently, they adopt a child," he explained. "Then the woman finally relaxes and she becomes pregnant."

I worried I was too old, and besides working at the local newspaper, I owned the Angler's Calendar Company. Did I have time to raise a child and work two jobs? Mike and I debated the future and decided a child would be fine. We'd adjust. Once my belly became enlarged, he became unusually protective. He hovered over me, attentive and restrictive. No longer could I snow ski, no longer could I carry suitcases when we traveled, and no longer could I lift heavy calendar boxes.

Yet, I could become extra weight on a tractor blade. Mike wanted a patio extending from the living room French doors and chose to smooth the dirt by himself. He drove a tractor from the ranch headquarters and attached fifty-pound weights to a scraping blade.

"Why don't you get on? You'll add another hundred pounds at least," he said. "I'll have this area smoothed in no time."

I mounted the blade and held on tight like a bird on a branch. If my parents could only see me now: seven months pregnant in a flowing red dress, standing on a blade behind the tractor. I held onto the back of the tractor seat as Mike propelled the machine around the yard. Before long the patio area was level, and Mike relieved me of my chore. Melvin and Mario arrived with wood planks to enclose the space and poured cement into the center for our future patio.

A few Saturdays later, we heard the whump, whump, whump of helicopter blades while arranging items in our three-vehicle shed. In the sky, Jack and Elaine Wright, owners of Kimberly Nursery, flew over our house. They looked like they were riding in a green dragonfly, turning left and right, as they surveyed our property. They landed in the field behind our house and stayed to pick out shrubs, trees, and irrigation sites for our landscaping project. Growing up in a New York suburb, I thought their arrival by helicopter seemed straight out of a scene from the television show *Dallas*. While they talked figures and locations, I left to make Mike's lunch: two tuna-fish sandwiches, a handful of potato chips, three glasses of milk, and five Oreo cookies. He never varied his meals; consequently, it was easy for me to shop and prepare. Before long I heard the helicopter leaving and Mike coming in through the garage door. His lunch waited for him on the kitchen table.

* * *

Eight and a half months after our marriage, I began to experience intense contractions. Merle Wolverton had died three weeks earlier and I initially thought my strong sensations were related to his recent funeral. I woke from a sound sleep with an odd sensation. My stomach felt upset, like I had eaten something bad. I went into the living room, dressed in my flannel nightgown, and lay on the couch, trying to relax. But I couldn't. I wandered back to the bedroom, holding my extended belly in my hands.

"Mike. I think I'm in labor," I said.

"No way. You've got another two weeks to go," he said. "It's just false labor. Come back to bed."

I crawled into bed but felt too uncomfortable to stay. Again, I departed for the living room. Struggling to ease the pains, I curled into a fetal position. Finally, I shifted to the floor and got on my hands and knees, assuming the yoga 'downward dog' position. I extended my butt in the air and my forearms on the floor, stretched out straight. I made it through the rest of the night, but as the sun peaked over the foothills, my contractions became harder and closer together. I called Dr. Miller.

"You live too far away. You better come in now," he instructed.

"Now?" I questioned. "But my due date isn't for another two weeks."

He snorted at my ignorance and informed me due dates are just suggested dates, not facts. I reiterated the news to Mike. He again asserted I was in false labor. An hour later and wracked with pain, I declared, "No more waiting! We have to leave now!"

Mike brushed his teeth, threw on some clothes, and helped me into his pickup, frowning and shaking his head with displeasure. He was obviously aware of birthing cattle, had two children from a previous marriage, and therefore dismissed my concern. He *knew* I was in false labor.

On our way to the hospital, we diverted to the *Times News*. A salesman came to Mike's truck and retrieved the ads from my open window. Whenever I experienced another contraction, I stopped explaining the material. The men stared at me, waiting for me to continue. Another spasm struck my body and a tiny form shifted inside me. I finished talking and told my co-worker I'd see him in a week or so. Little did I know this would be my last day at the newspaper. Raising a child turned out to be much more time consuming than I had ever imagined.

From there we drove to the hospital. Once at the front entrance, an attendant placed me in a wheelchair and rolled me into a sequestered room and helped me undress. She held my arm while I stepped onto a stool and turned to sit on a disposable paper sheet that covered the birthing table. Next, she placed my feet into metal stirrups, and a nurse arrived to examine my cervix.

"You're dilated to three. You're on your way!" she said as she gave my arm a cheerful pat and smiled at me.

On the labor table in a small cubicle, I found myself surrounded by fishermen: Dr. Miller; Dr. Lambert, a pediatrician; and Mike. While I lay covered with a cloth on the table, they discussed salmon and trout. Whenever I cried out in agony, all three stopped talking and waited until the "noise" had passed. Then they began their stories again. Because the hospital was under repair, the only thing separating me from strangers and family members was a thick curtain. These groups sat about thirty feet from me while I gasped for air, trying not to scream for fear the people on the other side of the heavy drape would hear me. I never expected such terrific pain and I shouted and cried during the last hour of labor.

"He's coming," Dr. Miller said. "I can see the top of his head. Bare down, Bobbi."

Sweat ran off my forehead as I tried to comply. My wet body soaked the paper sheets and I squeezed Mike's hand as I pushed. At 1:30 p.m. on Thursday, May 5, my son was born. So much for false labor. At the foot of the bed, Dr. Miller held the tiny form. Because Mike and I were completely unprepared, we had no name picked out, no baby supplies, and my mother's flight from Florida was not to arrive for another ten days.

A nurse cleaned the infant and placed his nude body on my stomach. I looked him over and saw he had all his fingers and toes and a shock of black hair. "So, that's what a newborn looks like," I said. She then wrapped him in a soft blanket and handed the bundle to Mike. He cradled the tiny infant in his large hands and left the room to show his family, waiting on the other side of the curtain. An attendant wheeled me upstairs and I immediately fell asleep. Mike arrived later with a green plant and a colorful bouquet of flowers—not only for the infant—but to wish me an early happy Mother's Day. I was now a mother. Who would have thought?

The following day, the three of us drove to Kimberly, visiting relatives and having dinner. Before long, I was drained. We drove east on Highway 30 until we hit 4900. From there it took only ten minutes to arrive at our home, and I could finally go to bed.

"Mike. Something's wrong," I said standing in the bedroom a few minutes later. "The baby doesn't seem right."

"What'd you mean? How would you know?" he answered while already under the bed covers. "You've never had a baby before."

"I don't know. But he just doesn't seem right. He seems exhausted," I said as I sat on the bed beside Mike.

"Of course, he does. He's had a busy day," Mike countered and turned over to go to sleep. I didn't move but continued to sit next to him, staring at his blanket-covered body.

"Shit! Okay. We'll take him back," he grumbled as he whipped off the covers and climbed out of bed.

After putting on my coat, I grabbed the diaper bag and swooped the baby from his crib. Mike waited for us in the pickup. Uncomfortable silence filled the truck during our hour return to the hospital. We entered the emergency entrance and a doctor took us into a small examining area. A nurse drew the infant's blood and stated the numbers to the doctor. They were excessive.

"Much higher and he could have had brain damage," the doctor told us. "He's very lucky you noticed his sluggishness. We'll keep him for a few days and put him under bilirubin lights. His numbers will come down, and he'll be safe to take home."

We were told that they would place the baby under special lights to combat infant jaundice. My instinct as a mother, even a brand-new mother, was correct. Three days later Mike and I brought our little bundle home from the hospital.

"Put him in sunlight for about fifteen minutes each day," Dr. Lambert instructed. "That will keep his numbers low and you won't have to worry about bringing him back to the hospital."

After we arrived home, I placed the nude infant on a white towel beneath our bedroom window. Direct sunlight covered him as he slept. Fifteen minutes later, I picked him up and laid him in my family's heirloom crib.

As the bedroom windows caught the morning sunlight the following day, I again positioned our baby on a towel in the bright glare of the sun before climbing the front stairs to my office. How peaceful he looked, I thought as I saw him lying in the golden rays shining through the windows. Being a career woman for most of my life, I immediately forgot about the child. Forty-five minutes later, I remembered my baby was still in the sun. I raced down the stairs to our bedroom. There on the floor lay a bright red blob—as red as a hot chili pepper and as motionless as could be.

My gosh! Did I kill him? I thought.

I carefully picked him up and rushed to the bathroom and filled the sink with cold water. He didn't move and I was scared to death. *What have I done?*

Once the sink was full, I plunged him into the chilly water. A piercing scream emitted from his little body. Yes! He was alive. He was probably unhappy, but he was definitely alive. Gently wrapping him in a towel, I held him close. Cradling him against my chest, I whispered that I loved him dearly. He would just have to have patience with me. I carried the tiny baby to the kitchen and called Mike on the ranch radio. "Mike, can you come home? Clear."

Because our radio transmitted to other ranches, no private conversations were had. I didn't want anyone to know what I had done. Understanding the many radio connections, Mike answered with a brief response.

"I'll be right there. Clear."

Although nothing terrible had happened, Mike calmed me and looked at his tiny boy. "I don't know why things seem to happen around you," Mike said as he wrapped his arm over my shoulder and pulled me closer. "But they do."

"Next time I'll set a timer. I won't forget him again," I promised Mike as we left the sleeping infant in his crib and walked to the kitchen.

Chapter Twelve

Angler's Calendar Company

I created the Angler's Calendar Company in 1975, four years before meeting Mike. My first Idaho employees, Diana Breeding and Karen Brown, came by word of mouth. But I needed more help. As I drove north on 4900 one day, I saw a woman mowing her lawn beside the road. I stopped the car and walked over to introduce myself.

"Hi. I'm Bobbi Wolverton. I live a few miles south of here," I said.

"I'm Cathy Humphries," she said as she bent over and wiped sweat from her face with the tail of her shirt.

"I just moved here and own a small business," I said. "Would you like to make some extra money and work for me?"

"What would I have to do?"

"Answer the phone, call retail stores, and package orders," I answered.

"How much do you pay?"

"I pay minimum but I give a pretty good bonus at the end of the year," I said. "And you can take time off for children's school or doctor's appointments."

"When would you need me?" she asked.

"As soon as possible."

"I'll be there tomorrow," Cathy said. And she was. And she continued to work at the Angler's office for the next twelve years. All because of our chance encounter while she mowed her lawn the day I happened to drive by.

The Angler's office was situated over the garage of our home with its own entrance and staircase. There were four desks with typewriters; a graphic light table for designing; file cabinets; and a free-standing, professional copy machine. The bloodline of the company was the telephone. With sixteen hundred stores that had to be contacted annually, employees were on the telephone almost all year long. However, the local telephone company allowed each family or business only one telephone line, which was also a party line. We were connected to three other families. Murtaugh numbers were only four digits; ours were 6611. Except for phoning Murtaugh, all telephone calls were billed as long-distance calls. Once school let out and students arrived home, we lost all communication. We telephoned nonstop during school hours and then completed paperwork the rest of the afternoon.

Sharon Kimber, who first started cleaning my house, was later hired to handle incoming supplies and outgoing orders. Eventually, she became the shipping manager. On many days, the complete staff contributed to packaging and shipping. It was too much for one person to handle. On one sales-breaking day, we sent over a hundred boxes and completely filled the UPS truck. To work from home was a true blessing, and my business soared thanks to my competent staff.

Chapter Thirteen

Failing Brakes

"Want to ride with me?" Mike asked over the radio. "I have to drive to Kimberly. Clear."

"Sure, I'd love to. When will you be here? Clear," I answered.

"Five minutes, clear."

He drove a ranch truck with a large, tightly packed trailer filled with red- and white-dotted beans, and pulled into our circular driveway. I bundled Matt, our four-month-old infant, into a flannel blanket and placed him on the floor of the truck. Wearing jeans and a short-sleeve shirt, I grabbed the handle near the front window and hoisted myself into the cab. Once I cradled Matt on my lap, we began the journey.

We turned west from our home before heading north on 4900. The clean air and dazzling sunlight made for a spectacular late-summer day. The four-mile drive was uneventful until we came over a small rise at the end of the road. As we started down the hill, Mike tapped the brakes. We were approaching Highway 30 and needed to halt at the cross road. Nothing happened as he attempted to stop the heavily loaded truck. He leaned back and slammed on the brakes. Nothing.

We continued toward the stop sign, at a speed way too fast to handle the corner. Mike shifted down and applied the hand brake.

The truck slowed but there was no way it could stop in time. Mike turned the steering wheel hard to the left while shifting again. We flew around the corner, tires squealing on the pavement.

"What happened?" I shouted over the roar of gravel slicing into the truck's underside as we decelerated and maneuvered away from the roadside borrow pit.

"The brakes failed," Mike said as he brought out a handkerchief from his back pocket and wiped his brow. "I still need to get rid of these beans. I'll drive super slow to Kimberly so I can unload them."

He lit a cigarette and took a deep drag. We both opened our windows, and Mike blew smoke from the cab. I let out a sigh of relief and Mike repeatedly pumped the brakes as we continued on our way. They seemed to be responding, but just a little.

"How's the baby?" he asked, glancing at our infant lying in my lap.

"Sleeping. He has no idea what we just went through. Will Terry be able to fix the truck?"

"Yup. No problem there," Mike answered. "He's one good mechanic."

A stand of silos on the outskirts of Kimberly bordered Highway 30. We passed an idle line of freight cars on the railroad tracks and edged into a space near the silos. Our truck crawled to its proper location; and Mike opened the back of the truck, allowing the pinto beans to spill out, sliding down a conveyer belt into a rectangle container. I held Matt in my arms and whispered soothing sounds as the beans filled the metal box. He looked up at me and yawned, sleepy from the half-hour drive. He pressed his soft lips into my shoulder and began to coo. I cradled his head next to my body and rocked him back to sleep.

Mike walked around the truck and inspected the wheels to see if there was anything he could do. He shook his head. This would definitely be a job for Terry. He wiped his hands with his ever-present, multi-tasking bandana and climbed into the cab. Within minutes we were on our way, slowly driving back to Sky Ranch.

Chapter Fourteen

Coyotes

During the evening at the ranch, we often heard howls coming from far away. I had no idea what the noise could be. Mike said, "Coyotes."

I noticed how differently he pronounced the word. Coming from the East, I pronounced it "Ki' yote tee," giving the word three syllables. In the West, they pronounce the word "Ki' yote" with only two syllables.

A coyote is a dog-like animal, smaller than a wolf but larger than a fox, and they roamed throughout Idaho. They are great rodent hunters but will attack anything easy enough to kill and eat: rabbits, opossums, racoons, porcupines, dogs, and cats. Their soft, thick, long-haired fur, often a mixed color of greys and tans, brought a good price to hunters as their winter fur is considered exceptionally luxurious in the fashion industry. Mike had killed six coyotes the year before; the money he received from the pelts paid for our California vacation. As there were no limits on pelts, Mike and his friends hunted as often as possible. I heard one of the hunters brought a recording of a rabbit being skinned alive. The screaming of the rabbit enticed a coyote to approach the sound. I hated that version of the hunt. I soon began to dislike all aspects of coyote hunting as I found the animals to be exceptionally beneficial to our ecosystem.

On one moonless night after Mike and Matt had already gone to bed, I heard a chorus of coyotes yowling in the distance, out in the field beyond our lawn. They had a yippee noise that went to a howl. Their voices mingled, rising into a crying crescendo, creating an auditory illusion of a large pack. In reality, there were probably only two or three animals. Their sounds filtered into the living room as I opened the French doors. Our new black Labrador surged outside, responding to the sound in the darkness. Sam had died two years before, and Jack had become his replacement. A ridge of hair stood up on his wide back. He looked over his shoulder to see if I were following and raced to investigate the sounds.

He came back after scaring the coyotes away, his hackles bristling. Then they started howling again. Jack stood motionless, one foreleg slightly raised, his eyes fixed on a point somewhere in the dark distance. He shot back, running full out with his head extended and his body bunching and flattening. He proceeded further into the field, his black body blending into the night darkness. Each time he returned, the howling stopped. He wagged his tail as if to say, "Look, Mom, they're gone."

Then the noise started again, and Jack turned to run back into the dark field, a low growl emerging from his throat, his ears crushed against his head. My eyes adjusted to the night as he flew from the back steps. This time when he returned, his red rag of a tongue protruded and droplets of saliva scattered about his jowls. Again, the howling started. I called to Jack and brought him into the house. I had surmised the coyotes' philosophy. They were wearing him down and would attack at his most vulnerable point, when he was absolutely exhausted and extremely weak. They were truly resourceful and had learned to exist in all sorts of environments, adapting their lives from woods to deserts and from farms to cities.

When Jack trotted into the kitchen, he headed for his water bowl. From there, he slowly walked toward the living room and flopped on his side next to the couch. For the rest of the night, he never moved again until the following morning. The coyotes had almost beaten him, but now he was inside and totally safe.

Chapter Fifteen

Spring Floods

A few months later, scattered clouds surrounded the nearby foothills. KMVT weather forecasters projected heavy, spring rains. Within an hour of their prediction, the skies opened up. A strong rainstorm fell to the foot-deep snow still lingering on the ground from earlier winter storms. Lightning flashed across the landscape, illuminating the fields behind our house. Loud thunder burst from the angry clouds and streaks flared again and again, painting the sky a deep purple. A massive deluge poured down, inundating Sky Ranch.

Mike and I sat watching television in the living room while our toddler played with green and red toy trucks on his quilted blanket. Huge raindrops splattered against our windows as the wind intensified. A strong, steady rattling beat. A moment passed. Then we heard a familiar, crackling sizzle, and the house was plunged into darkness.

"Damn. We've lost power. Again," Mike exclaimed.

As this was a regular occurrence during powerful storms at the ranch, we had lanterns and candles spread throughout the house. Holding a flashlight, I headed to the kitchen and reached for a glass lamp in the pantry. I placed the flashlight on a counter and struck a match, igniting the kerosene-soaked wick. Matt crawled after me as

I meandered through the living and dining rooms, lighting several lanterns and candles.

Once we had enough light to read by, I chose a colorful children's book and scooped Matt from the floor and placed him on my lap. Using the glow from a lantern with its wick turned high, I began to read a story. Mike sat in a wingback chair and reached into his shirt pocket and drew out a pack of Marlboros. Shaking a cigarette out, he placed it in his mouth and bent toward a candle. Once lit, he took a long drag and leaned back. Smoke rose above his head and the tip glowed a bright red as he glanced at the *Times News.*

Beams from candles flickered across the walls, casting eerie shadows as the storm raged outside. Gusts of wind hit our windows like waves in a storm surf. Then all was quiet as we waited for another round of thunder. Outdoors the wind blew from the west, and giant rain drops pelted our windows. Jack huddled close to my feet, shivering with fear; he hated thunderstorms. And yet when we duck hunted, gunshot sounds meant nothing to him.

"Did you know the Oakley dam is threatened?" Mike asked as he read the newspaper.

"No. What's going on?"

"With all this rain and with the snow melting so quickly, the reservoir's to capacity," he answered.

"What will happen?"

"They've added extra wood to the top of the dam."

"Will that hold back the water?" I asked.

"No. There's way too much water to stop the overflow. As soon as the skies clear, a bunch of farmers are going to dig a flood trench, moving the water away from Goose Creek. You know, the river that flows through Burley."

"I did hear about a possible flood. But I don't know any details. Are you going to help?"

"Of course," Mike said. "We should get it finished in a few days."

The following morning a bright sun finally shone down, and Mike left to help with the digging project. I climbed into my car and drove to a viewing site by the dam. Balancing Matt on my left hip, I exited the car and surveyed the scene from the top of a bluff that overlooked the giant gully below. I had brought an old bedspread and a picnic basket filled with tuna fish sandwiches, potato chips, cookies, and a thermos of milk. Mike jumped down from a road grader and climbed to the top of the ridge to join us for lunch. We sat in a half circle and watched dozens of farmers and their employees moving dirt. They had tractors, dump trucks, plows, backhoes, trucks, road graders, front-end loaders, mechanical shovels, and the ranch's Caterpillar bulldozer. Even a gasoline tanker roamed the bottom land, filling machines and saving time so everyone could continue working. From our high vantage point, it looked like children's Matchbox toys, vehicles in vivid colors, moving every which way. The machines shoveled dirt to each side and dug a large passage away from the original Goose Creek channel. A smaller ditch spread out toward Sky Ranch on a path to Murtaugh Lake. Many of the machines had bright headlights, allowing everyone to work in shifts through the night. The massive mission in 1984 was finished in a matter of days, just as Mike had predicted. I was impressed by the men, working side by side without pay to help with Burley's crisis, no matter their farming status, religion, or ethnic background.

Two days after that immense undertaking, I attended a philanthropic educational organization (P.E.O.) in Twin Falls. I bundled Matt in his car seat and met Georgina, my sister-in-law, at the women's gathering. She informed me that 4900 had been washed out because of the immense water overflowing from the Oakley Dam. On my way home that night, I decided to cut south near the Murtaugh Café on Highway 30 and avoid 4900. Once I circled the west end of the town's lake, I came around a curve and approached a breach in the road. A fast-moving river, several feet deep, rushed across it. The initial creek had swollen above its capacity and had eroded the road I had chosen. In the pitch-black night I stopped and stared. My headlights illuminated flashing white caps. A swift current tore from the right edge of the road and boiled over several protruding rocks. I debated the situation as the water plummeted down into a quagmire, a channel swirling and churning toward Murtaugh Lake.

Thinking my Subaru was invincible, I decided to cross. Inching the car forward, I steered to the center of the road. The rushing water rammed into the side of my car, sounding like a blast from an oncoming locomotive. I tightened my hands on the steering wheel as the water surged over the road and slammed my car sideways, toward a deep ditch on the lakeside. My heart thundered in my chest as waves overlapped the rocker panels and began to seep into the car. If the current had overtaken me, my Subaru would have plunged into the ditch and overturned. I would have been upside down with water flowing through the car. Terrified and shivering from fear, I continued. I clung to the steering wheel, my body hunched forward, and kept the Subaru moving through the wide rush of water.

Finally, I made it to firm asphalt. It was as if an angel had wrapped its wings around my car and protected us from the flood. I leaned my head back against the headrest and took a deep breath. I waited and tried to calm myself. I had been so accustomed to city life and modern infrastructure, I had never experienced washouts and didn't realize how dangerous they could be. And being single for so many years, I had completely forgotten the small bundle tucked in his child seat, sleeping in the back. Shivering from fear and stupidity, I headed to the Foothills Road and continued home.

* * *

The Angler's Company had expanded, and I had way too much work to accomplish. Plus, the Federation of Fly Fishers (FFF),[1] had

---

1   Renamed Fly Fishers International in 2017.

asked me to produce a book to use as a fundraiser. As I knew the printing industry, I took on the challenge and contacted thirty fly casters, tyers, authors, and artists who all agreed to donate their expertise. I even had President Jimmy Carter write the dedication. At the end of the book, a dozen fly-fishing manufacturers consented to full-page advertisements. These sponsors paid for the production, and the FFF made a one-hundred-percent profit on each sale. It was a time-consuming but rewarding challenge, and as a result, the organization presented me with the prestigious Arnold Gingrich Memorial Life Award, recognizing me for outstanding writing achievement about the sport of fly fishing.

Chapter Sixteen

Mormons

"Are you feeling sick?" I asked Kay, my neighbor and constant babysitter.

She sat in a wide comfy chair with two children nestled on each side of her, watching cartoons on television. As soon as I let go of Matt's chubby hand, he tottered to her chair and scrambled onto her lap. I stood beside them as they stared at the television. When I looked down at Kay's shoulder, I noticed she wore something under her sheer pink nightgown.

"No. Why'd you ask?"

"Well, you're wearing a T-shirt under your nightie," I said. "I thought you had a cold."

"That's my garment," she said looking at her shoulder. "If you're LDS, you wear the garment next to your body. It's part of our religion."

I didn't want to pry, so I said nothing more. But as soon as I returned to my office, I asked one of my employees about Mormon garments.

"It's what they wear once they've gone through a temple ritual. They wear them day and night unless they're bathing or swimming."

"What do they look like?" I asked.

"It's two white pieces: a top with capped sleeves and underpants that taper to their knees. Women wear the tops next to their skin, under their bras," she added.

"Why?"

"I'm not sure. I've heard it makes them feel safe. They're pretty secretive, so I really don't know."

For another hour we talked about Mormons, a shortened word to designate the people who belong to the Church of Jesus Christ of Latter-Day Saints (or LDS). Their churches are called "wards," and except for family gatherings, most of their socializing is done at the wards.

Thirty miles east of Sky Ranch, a "beehive" farm existed. Whenever I'd drive to Burley, I'd pass a large white sign depicting a brown beehive with honey bees flying around it. The farmland surrounding the sign was owned by the Mormon Church and most all its workers were church volunteers. The food raised would be given to needy church members and to disaster victims around the United States. They believed in taking care of their own and helping others whenever possible.

I also discovered that Mormons kept a year's supply of food and water in their homes. Because most of my neighbors belonged to the LDS Church, I began to follow their principles. I might not have had a year's supply, but I certainly had enough food and water for a few months. Another interesting LDS assignment was Family Home Evening. It was a special time for games and spiritual guidance. Matt loved to go to the Nebeker's on Monday night and play games with their many children. I sometimes joined the family in their fun-filled evening activities. As Monday approached, the designated child would plan the night's entertainment. Sometimes it was a play, sometimes a mini cross-country event, sometimes hide-and-seek.

Living in a Mormon community, I learned about their tithing, missionary work, and baptism ceremonies. For tithing, they gave ten percent of the income to the Church—and that was before taxes. Missionary work was achieved by Mormons moving to different states and countries and proselytizing to the locals. The men went for two years and the women for eighteen months. Surprisingly, their time on a mission was paid by their family and friends. This meant the missionaries were responsible for the expense of their transportation, lodging, and food. If they needed to learn a language, they moved to Provo, Utah, for nine weeks. Their language school is considered one of the best in the world and is a prototype for our own government's international employees. Whenever I saw a young man wearing a white shirt with a name plate over the left pocket, a dark tie, black slacks, and sometimes a dark suit jacket, I'd recognize him as a Mormon missionary.

An acquaintance of mine drove once a month to the temple in Salt Lake City. As a member of the LDS church, she would act as a stand-in during full-immersion baptisms, most of the time for individuals who had died many years before. The Mormons believed that to go to heaven, one had to be baptized. Consequently, she might be baptized a dozen or more times to help those who had not been baptized during their own lifetime.

Because Mormons do not drink alcohol or any caffeinated beverages, nor do they smoke, I found them to be a handsome group of people. Their skin does not wrinkle as much as others, and they basically seemed much healthier. They valued their families and worked hard both in their employment and in their home life. Although I did not share their religious beliefs, I was glad to have several Mormons as my good friends.

Chapter Seventeen

Tent Caterpillars
and Bathtub Living

At the edge of our lawn stood a sentry of aspen trees, marking the beginning of a potato field. Inside the house, our two-year-old son sat in his wooden highchair while I fed him raisin oatmeal. He wore a blue, one-piece romper. It was an early spring day, bright and clear, and unusually warm from the Chinook winds blowing down from Canada.

Beyond our dining room window, I saw numerous white caterpillar tents, thick gauze masses the size of dinner plates. They covered the lower limbs of the budding aspen trees, their leaves so thin that light shined right through them. These hungry insects were known to consume every leaf, damaging and sometimes killing the host tree.

After breakfast, I wiped Matt's tray and lifted him from the chair. With his chubby little hand in mine, we walked over the gravel driveway to a metal shed near the back of our house. From the dark recesses of the three-vehicle building, I retrieved a large branch pruner and a red gasoline container.

"Let's get those nasty caterpillars," I said.

With imminent destruction in mind, I marched to the trees. Matt followed close behind, and I began lopping off tent-covered branches and throwing them into a pile. When I finished, the mound topped three feet. I sprayed gasoline over the tree limbs, interspersed with caterpillar tents. With concern for safety, I moved the red can thirty feet away and turned on our green garden hose, dragging it from the house to be close to the branches. I made a little spot at the base of the mound, preparing to light the tree limbs.

"Matt, stay behind me. Don't get any closer," I warned.

He moved as I had instructed and watched over my shoulder when I knelt to ignite the branches. I struck the wood match to the rough side of the box and a flame instantly blazed. With nothing but a quick brush under the branches, I placed the burning end to the base of the mound.

A roar ripped through the air and a huge fireball exploded. The match had ignited the gasoline and an enormous blaze blasted over my head. A gigantic flash flew upward and then curled down over my crouched body. I screamed and immediately dropped to the lawn and rolled, hoping to extinguish the supposed blaze on my nightie. Matt, still behind me, laughed out loud and clapped, jumping up and down with a huge smile on his face as I twisted on the grass.

"Do it again, Mommy!" he yelled.

Once I stood, I realized my nightgown had not even been touched. Matt, however, had received the brunt of the explosion. The inferno had spiraled over my bent body and torched his upper face, blackening his forehead. His long, brown bangs and eyebrows had almost disappeared in the flames.

"Oh, my gosh. What happened?"

I knelt and brushed my hand over Matt's hair, tangled in scorched knots, and examined his head, body, and pajamas. Picking him up, I wrapped him in my arms. Standing a few feet away, I held him close to my chest while the blazing fire consumed the remaining pile of branches.

"Do you feel okay?" I asked. "Are you hurt?"

Matt was not at all concerned and stared at the flames while high in my arms, smiling as if we were at a birthday celebration.

"I'm calling your Dad," I said to Matt. "I want him to come home."

It took a few minutes for the fire to diminish. I showered the ashes with water, walked back to the house with Matt in my arms, and turned off the hose. Inside the kitchen, I put Matt down and reached for the handheld radio to call his father.

"Mike. Can you come home? Clear," I said.

"I'll be right there. Clear." he said.

Just then the phone rang.

"Is Matt all right?" my mother-in-law asked. She had heard my message on the ranch radio. She didn't worry about me, just about her grandson. She knew things happened around me and she wanted to be sure he was safe. She often mentioned that Matt would be lucky to survive his mother.

"Yes. He's okay. Mike just pulled in. I'll call you back in a few minutes," I said.

"What happened?" Mike asked as he walked in the back door. His blue eyes flashed from beneath his baseball cap as he spoke.

I looked up to meet Mike's steady gaze and told him about the early morning effort to remove the tent caterpillars and the ensuing fire. As he listened, he took off his cap and ran his hand through his blond hair, staring at me in disbelief. He bent over and checked Matt's head and acknowledged he was fine. There was nothing to worry about.

"Don't you know gasoline is flammable?" he asked in an angry voice, a look of disdain on his tanned face. A vein in his neck bulged as he forced his question to me.

"Of course I know it's flammable. That's why I used it." I countered. "I didn't know it was explosive!"

After my comment, Mike shook his head in disgust and walked down the hall toward his office. I felt chastised and wondered why he was so critical of me. Was I too sensitive? Or was he too harsh?

I changed Matt's clothes and replaced my nightie with jeans and a shirt. The challenges of living on the ranch sometimes seemed overwhelming. But life on the ranch was exactly what I wanted. The peace and beauty were perfect for me. My transition from city gal to country wife was far from complete but still stimulating and exciting.

* * *

When we designed our home, we didn't know I could become pregnant. Consequently, the master bedroom was at the far end of the house on the first floor. Once Matt outgrew his crib, I didn't feel comfortable having him so far away, in a bedroom on the second floor. As our downstairs guest bathroom was close to the master suite, I chose that space for our son. I placed a small foam mattress in the tub and covered it with sheets and blankets. The bathtub became his bed. I removed the shower curtain and hung his clothes on the rod above the tub. The counter top held a thick changing pad, and the drawers contained all his baby supplies. Little Matt now had a complete bedroom, right next to ours.

For the next two years, he slept in the tub. That situation came to an end once I hired a city plumber to fix a leak in the tub's faucet. He kept looking at me and then back to the bed in the tub. I thought he'd report me to child welfare. That very night I removed Matt's items and shifted them to a bedroom on the second floor.

Chapter Eighteen

Airplane Travel

When I arrived in Idaho in 1980, the state's population was under a million inhabitants and would not break through that statistical barrier for another ten years. Idaho is a big state with few people. Consequently, I loved the fresh air and clean water; and with such a small population in Magic Valley, I rarely had to wait in line at grocery stores or movie theatres.

The negative aspect of living in a sparsely populated area was having to drive long distances or travel by air on a small, thirteen-seater prop plane. Because my family lived back east, we flew from Joslin Field, now called Magic Valley Regional Airport, to visit them in Florida. The rectangular building, located five miles south of Twin Falls, supported only a few flights a day, all of them going first to Salt Lake City. Passengers then transferred to larger planes to continue their journey across country.

On one of our trips to the East Coast, Mike, three-year-old Matt, and I flew from Joslin Field in a fierce blizzard. Strong winds battered cars and buildings as we drove along the Foothills Road toward the airport. In one extreme gust, we later heard the historic Hollifield barn, built in the 1910s in Hansen, had blown off its foundation.

The blistering snowstorm delayed the arrival of our airplane. While waiting we checked out a huge wood-carved mural that covered one wall and ceiling of the airport hallway. Kimberly artist Gary Stone depicted numerous scenes of Magic Valley, most of which we recognized. Because he continued almost daily with his art, adding more and more businesses, we decided we'd have him carve and paint a fly rod and creel to represent us. He and his wife, Bev, had been my friends for years, and I had edited their Christmas book, *The Secret of Santa Claus*. From the hallway, we walked to the adjoining restaurant and sat at a small table, eating donuts and drinking coffee and hot chocolate, waiting for the plane to land.

"The flight to Salt Lake City will begin to board in fifteen minutes," an employee announced over the loudspeaker.

"Let's use the bathroom," I suggested. "It's going to be a rough flight."

Once the plane landed, we bundled in thick jackets, covered our heads with caps and hoods, and raced from the terminal to the portable stairs while icy snow pelted our bodies. The silver aircraft had a center aisle with a seat on each side, extending five rows back from the cockpit. Three seats across the rear completed the cabin layout. Once inside, we placed our jackets in the overhead compartments and parked ourselves in the middle of the aircraft. Matt and Mike settled in hard plastic seats across from each other and I sat in front of Matt on the left-hand side. We tightly buckled ourselves and listened to the safety instructions from a female attendant as we traveled down the taxiway. We didn't know it then but a frightening ordeal had just begun.

As a former flight attendant, I was used to difficult flights in all types of weather. I had worked for an international airline and thought nothing could make me queasy. That was until I took this flight. Passage over the Albion Mountains in the middle of a blizzard is not something any airplane should attempt to do. The passengers should have cancelled their flights. I ought to have known better.

When the small plane flew upward to cross the mountain range, it twisted and turned with every blast of wind. Although it was nine in the morning, we saw nothing from our windows. The sky was black with dark, snow-laden clouds. The plane continued to spiral, shifting and diving. We were thrown against our seat belts like rag dolls. Soon an unsettling feeling in my stomach overtook me. But I was determined to make it to the airport. I was not going to get sick. Then we heard an announcement over the intercom.

"Ladies and Gentlemen, this is the captain speaking. Salt Lake Airport has temporarily closed. We have enough fuel to circle for forty minutes. If it doesn't open by then, we'll return to Twin Falls."

That was it. There was no way I could hold my upset stomach any longer. I reached for a plastic bag in the front pocket. Bending over, I retched into the container.

"Mommy. I want to sit with you," Matt cried as he unbuckled his belt and started to climb out of his seat.

"No, Matt," Mike said as he stopped our toddler. "Mommy's busy right now. Come to me." Mike lifted Matt and placed him on his lap, holding him tight while the plane jolted up and down. Matt faced Mike and started to cry.

"Oh, no," I heard Mike say while my head was bent over the burp bag. I turned around to see Matt vomiting hot chocolate over the front of Mike's sport jacket. I couldn't help it. I broke into laugher. My weird sense of humor certainly did not endear me to my husband. The man across the aisle also started to snicker. Maybe it was the stress of the flight that caused us to laugh, like stifling a giggle when listening to a funeral sermon.

As we started to descend into Salt Lake City, it was as if we were inside a toy slinky, rotating and twisting as we fell from the clouds. The man in the aisle across from me held his arm rests with white fingers and closed his eyes as we bounced across the sky. Finally, another message was announced.

"Ladies and Gentlemen. The Salt Lake airport has just opened."

No one clapped with gratitude. We were holding our personal barf bags, and vomit smells permeated the aircraft. When we landed and all the passengers had departed, I stood and gathered Matt's small backpack from under my seat. I changed his clothes and wrapped the soiled ones in a plastic bag and deposited them into his little pack. Mike had to stay in his smelly shirt and sport jacket for the continuing flight to Florida. In the Salt Lake airport, I took his blazer into the ladies' room and cleaned it as best as I could. Mike wiped off the worst of his shirt in the men's room, but it was no use. Once we boarded the plane heading to Miami, no one wanted to sit near us. The atrocious smell meant we had two rows of airplane seats to ourselves . . . the only blessing after our horrendous flight from Twin Falls.

After ten days visiting my parents as well as fishing and photographing the coastal region, we returned to Sky Ranch. Spring soon enveloped South Central Idaho; and Mike and Don, once again, began their farming chores.

Chapter Nineteen

Losing Our Toddler

"**M**att! Where are you?" I called, my voice slicing through the air.

I had started to set the table for lunch and stepped outside to pick some pansies, barely peeking their heads above the bits of remaining snow near our house. Matt usually followed me everywhere; now he was nowhere to be seen.

Huge piles of dirty snow lingered beside our dirt road. As the sun warmed, the leftover winter snow melted. The resulting water created a small pond about two feet deep a hundred yards or so from our lawn. Nine months of the year, this area was dry but after a quick snow melt and a wet spring, it could be particularly hazardous for a little toddler. I began to fear the worst for Matt. After checking inside and outside our house and in our barn and shed, I ran into the house and tracked Mike in his office, relaxing in his leather recliner.

"I can't find Matt. I've looked everywhere."

Mike thought the same thing I did. We were both worried about the nearby pond. He jumped from his chair and ran outside, losing a slipper in the remaining snow as he raced toward the pond. I was right behind him and waded into mud, then into deeper sections, my feet wrapped in freezing cold water. I imagined all sorts of horrible things and tried to remember the many stages of artificial resuscitation from my earlier days as a flight attendant.

Just then we both saw Matt. He bounded down the dirt road bordering the pond, his small fist buried around Jack's red collar. What a sight. He was muddied from his knees to his feet, and yet he had the biggest smile covering his face. Mike scooped him up and gave him a swat on his butt while hugging him close to his chest.

"Don't ever leave again without telling your Mom or me where you're going," he lectured.

Tears escaped Matt's eyes and covered his splotchy face. Farming is dangerous. Because workers are in close contact with gigantic machinery, fast-moving equipment, and large animals, there are many accidents and numerous fatalities each year. Being a wife and mother, I was always concerned about my family's welfare. Having a small child walking around ranch roads only added to my stress. Thank goodness, we only lost Matt two times during his early growing years. Yet with each occasion, Mike and I experienced major panic, always expecting the worse.

* * *

"Are you making something for P.E.O.?" Georgina asked over the phone as she prepared for the educational meeting.

"Yes. I'm cooking brownies," I answered. "But would you mind picking them up? I can't make today's meeting." I could prepare brownies blindfolded as the ingredients were in a box, and all I had to do was add eggs, oil, and water. Easy. Georgina arrived an hour later and walked into an acrid-smelling kitchen. She saw me hunkered over my batch of brownies, cutting off the burnt bottoms.

"Oh, no. You don't want me to bring those, do you?" She asked. "They'll think they're mine!"

"I'll put them on a nice plate and cover them with aluminum foil," I countered. "They'll never know. You can leave the plate as a donation."

She left the house with a disgruntled look on her face. I had burnt so many items during those first years that the smoke alarm became my timer. From an early age, Matt took a towel whenever the smoke alarm went off and whipped it around the air, helping me clear out the smoke.

*   *   *

After the brownie incident, I remembered the time a fire could have been disastrous. Mike was in his office, tying fishing flies, and Matt, at this time only a few months old, was in the living room inside a playpen. I was about to fry potatoes in oil and had set the burner on high. Just then Mike called me.

"Come see this fly I'm tying," Mike shouted. I forgot all about the pan of grease and walked into his office. Leaning over a Renzetti vice at the edge of his desk, I inspected the new fishing fly through a magnifying glass. It was an involved salmon fly with many feathers in several bright colors. He swiveled back in his oak chair, proud of his accomplishment, and we started to chat.

Soon Matt began to cry. I left Mike and walked back to the living room. From the hallway, I saw flames two feet high, coming from the pan of grease in the kitchen. Yelling to Mike, I raced to the stove and grabbed the handle of the pan with a pot holder. Knowing not to put it in water, I placed the blazing pan on our Corian counter top and threw flour on the flames, dousing the fire. When I raised the pan, our counter had a black circle beneath it. Dinner was put on hold while I checked on Matt. The next day I sanded off the scorched counter which created a slight indentation next to the kitchen sink. It became another reminder of my culinary calamities.

Although cooking was not my forte, I relished almost all aspects of business. With innovative concepts, my business expanded and won several awards, the most impressive being "Exporter of the Year for the State of Idaho" in the small-business category. Having a professional office in the house meant I could welcome Matt home from school, prepare meals easily for Mike, and handle international orders coming in at all times of the day.

Chapter Twenty

My Horse Smoky

"Want to sell your horse?" I asked my neighbor, Larry Adams, over the phone.

"You mean Smoky? I've never thought of it. You know he's twenty-two, don't you?" he stated. "If you still want him, let me know."

"I had no idea he was that old. I'll have our vet check him out," I replied.

In years past, I discovered Mike had participated in a rodeo event called "team roping." Larry Adams was the other half of the team. As Mike explained the event to me, Larry would ride his horse, Smoky, and lasso a steer's head, preferably around the horns. He and his horse were known as "headers." Mike then had the tougher job of roping one or both of the hind feet. He and Shamrock were known as "heelers." Once the ropes were secured around the head and hind hoofs, their horses backed up, tightening the rope and stretching the steer. Timing started the moment the steer and horses broke from their chutes. It ended with the animal stretched and immobile.

Before moving to Sky Ranch, I hadn't a clue about team roping. I had only been to one rodeo in my life and that was at Madison Square Garden in New York City when I was about ten.

Since I'd been riding since I was five, I wanted a horse of my own. With that in mind I rummaged through classified ads and talked about buying a horse to anyone who would listen. On my trips to town, I'd see Larry's beautiful black horse grazing in a nearby pasture along with a few other horses and several cows. As I passed the field on 4900, I saw that Smoky's long tail almost touched the ground. He looked like he was not only a Quarter horse as Larry had indicated, but also part Saddlebred as his mane and tail were so long and thick. He stood as if carved from ebony among the other animals, none of them with any shelter no matter the weather: pouring rain, blowing snow, or blistering heat.

I told Mike of my conversation with Larry, and that I wanted to have a horse, whether it was Smoky or not. We had the space and I loved to ride.

"Okay," he said. "I'll have Mario and Melvin construct an electric fence around the back."

Besides two acres of lawn, we had five acres behind our house. There would be plenty of land for one horse to roam. Since the ranch used center pivots to irrigate most of its land, each corner of a field covered seven acres. Sometimes a ranch family built a house on the corner, sometimes it would be put into alfalfa, and sometimes it was left unseeded, or fallow. Two men, along with Terry, had already constructed a small barn. The slate-colored enclosure had a box stall, inside lights, a heated water trough, and an outside light illuminating the entrance. Dirt covered the floor and a four-foot-wide opening allowed for easy access. Any horse I bought would be able to walk in or out as it pleased.

"Hey, Dean," Mike called on our landline. "Want to help us catch Larry's horse? Bobbi wants to buy it."

"Sure. I'll be right over."

Mike fetched Shamrock's lead rope, and the three of us ventured to Smoky's field a half mile away. Mike left his truck and ducked under the barbed-wire fence while Dean walked to the gate. With a handful of oats in a plastic bucket, Mike shook the container and called to Smoky. Being curious and loving oats, Smoky walked right to him. As the black horse licked the bottom of the bucket, Mike attached the lead rope to Smoky's leather halter and led him through the metal gate. I shifted to the driver's seat while Mike and Dean planted themselves on the open tailgate, their long legs swinging below. Smoky stood patiently behind the pickup, waiting to respond to whatever happened next. I drove slowly to our house with Smoky following at a trot.

Mike walked Smoky to the field and let him loose while I retrieved my camera. Smoky ran back and forth across the front of his enclosure, neighing and prancing. He had left the other horses in our neighbor's field and felt unsure of his new surroundings. While Mike and Dean surveyed Smoky, I photographed him. He held his head high, his mane flew upward in the wind, and his tail soared out behind.

"That's one beautiful horse," Dean exclaimed. "Glad I could help." He turned and climbed into his pickup.

Dr. Bob Monroe arrived that afternoon in his white veterinarian vehicle, strewn with medical supplies. Bob stepped from his truck and joined me in the barn to examine Smoky. He checked his teeth, nose, and eyes; tested his feet; ran his hand over his spine and down his legs; and took his temperature.

"Walk him around for a bit," he requested. "Then have him trot."

I held onto the lead rope, led Smoky outside the barn, and did as he asked. After directing Smoky to walk and trot in a small cir-cle, I brought him back to face Dr. Monroe.

"I think you've got a good horse there," he said. "But if he stayed in your neighbor's field without any shelter, he wouldn't live another two years. With a barn and good food, he should give you at least eight more years."

Before he left, he gave me a piece of advice. "Never leave him with a halter on when he's out in the field. I've seen too many horses get their hoofs caught when they're scratching their faces. That can be damaging and even life threatening."

After Bob disappeared down our dirt road, I practically danced to the house. I immediately telephoned Larry Adams and told him I'd bring a check over in a few minutes. I was so excited, I couldn't stand it. Once I returned from Larry's house, I called Mike on the radio.

"Mike. Smoky's ours," I shouted with glee. "Please thank the guys for their help with the barn and pasture. Clear."

I snickered to myself as I strode to the barn, skipping and laughing before I ducked under the hot-wired fence. Although I had owned a horse for a summer in high school, this was different. I now had a beautiful horse trotting around our pasture, one that I would have for years. Throughout the small barn, I arranged all my horse products. One basket held rags and another basket held a rubber curry comb, a stiff bristled brush, and a metal mane and tail comb. I then moved to the tack room within our shed and hung an English bridle, my half chaps, and a crop. The English saddle I had purchased from Vickers Western Store in Twin Falls already sat on a V-shaped mount, extending two feet from the wall and covered with a cotton dust cloth. Everything was in order and ready for Smoky.

After I had jogged back to the pasture and Smoky had calmed, I steered him to the side of the barn where Mario had dug a hole and inserted an eight-foot-high pole in cement. I attached his halter to a nylon cord, dangling from the post. His chin whiskers tickled my fingers when I gave him a carrot. He stood patiently while I brushed him and attempted to untangle his mane.

"Such a good boy," I murmured as I touched him and moved around the large gelding, always keeping my hands on his body so he'd know my location. "I'm going to take such good care of you. You'll love it here."

After a thorough cleaning, I led him back to his pasture. While I had been brushing Smoky, Mario had brought in a tractor with

its front scoop filled with a large bale of hay. The rectangular bale was four feet wide, four feet high, and eight feet long. It weighed about a ton. I held out another carrot and Smoky stretched his neck and gingerly took it. I rubbed behind his ears and he made a funny humming noise, his lips quivering in what looked like a smile. He seemed to be absolutely content, purring like a cat.

In all my excitement, I had completely forgotten dinner. I had been too thrilled with Smoky, the barn, and the tack room in our shed. Luckily, I could always fall back to salad, crackers, and a thick bowl of spicy chili, one of Mike's favorites.

After a few days had passed, I called my nearest neighbor, Marsha, on our landline.

"Want to go riding?" I asked as I stood in the kitchen, looking out at Smoky in the pasture behind our house.

"How about Saturday afternoon? My chores will be finished, and the kids will be in Twin with their Dad."

"Great. When and where?"

"Let's meet between our houses, about two o'clock. We'll ride to the gravel pit," Marsha said.

And so we did. After lunch on Saturday I gathered Smoky from his pasture and walked him to the barn. Once inside, I pushed the metal bit into his mouth, checked the cheek straps, and pulled the leather headpiece over his ears. *What a fine-looking horse*, I thought as I gripped his forelock and twisted it over the browband. Once Smoky was bridled, I threw a saddle blanket and English saddle over his back and bent low to grab the girth strap and cinched it to his saddle. Before venturing to Marsha's house, I rode him around his pasture, removing any quirks from his temperament. As we cantered down the dirt road, blowing tumbleweeds crossed our path and spooked Smoky, causing him to dance and twist. Marsha waited for me as I reached the halfway point.

"What's this?" Marsha said as she laughed at my English riding outfit and black helmet.

"I'm so glad you're not a wannabe cowgirl," she remarked. "There's no mistaking you for a city slicker."

"I've been riding since kindergarten," I stated. "I feel a lot more comfortable with an English saddle and bridle. The helmet is used for safety. It's the way I was raised."

"Good for you. I've never seen anyone riding like that out on the range," Marsha said. "Obviously, you're doing fine."

We cantered down the rest of the road with dust puffing out behind us and turned left and trotted up a rise on the rutted trail. The gravel pit extended north from her house, between BLM land and the Moss farm. We threaded our way into the depression and emerged on the other side, surging forward and loping along a path deep into government land. As far as we could see within our remote valley were tawny cheat grasses and teal sagebrush. The land seemed to go on indefinitely. Not another tree or any water in sight. It was void of roads and we melted into the sandy-colored basin. After dismounting, we walked in the sunshine among the different grasses with our horses trailing behind. As we moved further into the prairie, we stopped and listened. There were no sounds, no vehicles, no birds chirping, and no animals howling. This was the silence of the land. A sweet scent of sage sprang upward. I bent and looked at the subtle prairie. The closer you were to the land, the more you saw and smelled. Beautiful blue sky with puffs of white clouds floated lazily east toward the mountains. This was freedom. Not a care in the world.

"If we continued another couple of miles, we'd spot parts of the Oregon Trail. You can see a few deep grooves in the rocks near Milner. Thousands of wagons passed this way," Marsha said as we remounted.

"I was told you can also see wagon ruts near Murtaugh Lake," I said. "Especially in the winter after harvest. Amazing."

As shadows fell and the sun began to set, Marsha said, "Time to turn back. I have dinner to make and I haven't even started."

"Oh, dinner. That nasty word," I said as we turned our horses toward home.

* * *

From the day I bought Smoky, I switched him from Western tack to an English saddle and bridle. Instead of neck reining, as a rider does with Western tack, I used separate reins to turn Smoky's head. English riding also means posting up and down while the horse trots instead of sitting and bouncing in a Western saddle.

And I switched his riding style to comply with what I knew: figure eights and jumping. I put on work gloves and used a wheelbarrow to bring in a few dozen cinder blocks to create a small riding ring. The concrete chunks were only eight inches high, but it gave Smoky an area to limit his paces. At least twice a week, I rode him around our end of the ranch and then into the ring. By shifting my weight and pressing his sides with my knees, he learned to walk, trot, and canter in a figure eight within the cinder-block circle.

During a subsequent month, I decided to try jumping Smoky. I took six logs and laid them on the ground about two feet apart within the ring. I mounted Smoky and asked him to walk over them. He had no trouble and his feet never touched any of the logs. I next placed some cinder blocks on their sides and raised the logs eight inches. Again, he did fine. I moved the cinder blocks to a standing position, like a squad of soldiers, and put a log on top. Smoky easily trotted over it. He had handled everything perfectly so far. I thought, *Why not?*

Christa, Mike's daughter, played in the house with Matt while I rode around the ring. Oh, how I wished she and Matt had looked out the window when I attempted to have Smoky jump a fence. I dismounted Smoky and made a barrier about two feet high with an assortment of cinder blocks and two horizontal logs. Taking more logs, I created wings on the side of the hurdle, just like they do at professional horse shows. Now that everything was to my satisfaction, I put my foot in the stirrup and pulled myself up. He quivered with excitement as I arranged the reins. Once seated properly, I cantered Smoky slowly around the perimeter of the ring and then turned him to the middle. We were a team. Both of us moving as one. As we came to the hurdle, I cracked the crop on his shoulder and leaned forward, my hands on his neck.

Just as we approached the hurdle, Smoky abruptly stopped. I, however, flew skyward. My body performed a complete somersault, my feet snapped over my head and my legs stretched straight into the air. I plunged downward and touched the ground with my feet together, exactly like an Olympic gymnast. Smoky landed right beside me. He had decided to take the hurdle a nanosecond after he had unexpectedly stopped. We stood next to each other, looking into each other's eyes, gasping and puffing.

"Gosh, Smoky, two old bats and we sailed over the fence just fine," I said. "Next time, let's do it together." Smoky was not going to get away with that maneuver again. After remounting, we walked around the perimeter and both calmed down. Once we had settled, I had him canter the circle again and turned him into the jump. Just before we came to the hurdle, I gave him a sharp crack on his flank and kicked both sides with my heels. I felt the heave of his muscles and Smoky shot forward, surging toward the horizontal bar. He flew over the jump with room to spare. Patting his neck, I released the reins and let him leisurely walk around the circle.

"What a good boy, you are," I said. "I am so proud of you. Yes! I knew it. We're definitely a great team."

Chapter Twenty-One

Grasshopper Infestation and Hay Fires

"What's going on outside?" I asked Mike.

"It's grasshoppers and they're ruining our fields."

We had never seen anything like it. It was 1985, and millions of grasshoppers were inundating Magic Valley. The flying swarms ate crops by the hundreds of acres, chewing their way from field to field. It reminded me of the Biblical locust invasion. Eventually, the government stepped in. Idaho State standards specified a dangerous level when eight grasshoppers arrived per square yard. The newspaper reported there were over two hundred grasshoppers per square yard!

You couldn't avoid them when driving. They covered the road. One afternoon I ran a few errands and drove from Sky Ranch to Murtaugh, picking up supplies at Mark and Barb's Grocery Store. As I moved along the paved road, my car crushed their hard, shell-like bodies. The noise loudly reverberated and sounded like cracking china plates as I maneuvered around the worst of the green-brown piles of carcasses.

On the return, the squished bodies across the pavement become slick and slimy. My vehicle actually slid across one section of the road. As I drove down 4900, I saw a huge military plane fly close to the ground over nearby fields. Its camouflaged fuselage looked like a flying dinosaur while it sprayed killing chemicals over the farm lands. Ultimately, the infestation abated, and Sky Ranch continued to endure. Its crops had survived.

* * *

"There's a hay fire on 4900," I heard someone call over the ranch radio. I listened in the kitchen to what our farm employees were saying.

"The hay stack is about a mile from the highway. I think it's spontaneous combustion. Clear."

I looked out the front door but could see nothing to the north, not even a smoke trail. Nevertheless, I was curious.

"Hey, Matt. Want to go see a fire?" I asked.

"Where?"

"Not sure, but it's on 4900," I said. "Let's check it out."

I bundled our four-year old into his car seat, tightened the straps, and walked around to the driver's door. After snapping on my seatbelt and backing away from the garage, I headed out our driveway and started north on 4900. Then I saw dark smoke rising in the distance. As I drove closer, I saw an orange glow, ominous and threatening. I turned to the left and approached the giant inferno from a different direction. Sparks from a large haystack blasted into the air. It looked like the farmer was letting the fire burn out. There were no other vehicles nearby and the stack was in a plowed field, not close to any structures.

Once Matt and I advanced toward the burning hay, hot air struck the car windows. We were some eighty feet from the compressed stack, yet we could feel scorching heat inside the car. I backed up, not wanting to get any closer as flames shot upward from the middle of the bales. After a few minutes of staring and commenting, we drove home.

As Mike walked into the kitchen that evening, Matt ran to him, wearing his toy fireman's helmet, its siren blaring. With his arms outstretched, Mike reached down and lifted him above his shoulders.

"What's up?" he asked. "Has Mom been cooking?"

"No," I said and scrunched my face toward him. "Why does every fire mean I'm involved?"

"Well, look at the odds," Mike countered as he placed Matt back on the floor and took out a cigarette.

"No. I had nothing to do with this. It's about the hay fire on 4900," I said while scrubbing a few potatoes for dinner.

"I heard about that. They said it was spontaneous combustion," he stated as he drew in some cigarette smoke, slowly exhaling it upward.

"What does that mean?" I asked.

Mike walked to the kitchen table and sat down. I brought an ashtray and sat across from him. Matt raced between the living room and the kitchen, still wearing his fireman's helmet. Then he stopped to hear his father's explanation and turned off the siren.

"They probably cut the hay when it was too wet. After a rain, we'd roll over the rows of alfalfa, letting air dry the hay. Then we'd bale it. They might have baled the hay too quickly, before it was completely dry."

"Why would wet hay cause a fire?" I asked.

"Wet hay absorbs more moisture than dry hay," he explained. "Water is made of oxygen. You know, H2O. Oxygen is flammable. Hay acts as an insulator, causing heat to rise within the bale. Then a fire starts."

"Has Sky Ranch ever had one?"

"No. We've been lucky. But we've also made sure the hay is completely dry before it's baled."

"Did you hear what Dad said, Matt?" I said. "Another lesson for us."

I stood and checked our dinner. I pierced the potatoes to check their firmness. Nope. It'd be another half hour, giving me plenty of time to prepare a salad and broil two large steaks.

Chapter Twenty-Two

Halloween Snowstorm

Winter settled in with a vengeance, and for several days we were cooped inside, not venturing out except for me to check on Smoky. Although nothing much happened at Sky Ranch, the Angler's Calendar and Catalog business was in full swing. October, November, and December were our busiest months as orders from my catalog poured into the office. The full-color, forty-eight-page catalog had been sent to two hundred thousand individuals worldwide, displaying over a hundred high-end, fly-fishing gifts.

The Angler staff plowed through drifts of snow to come to work that October. I only had to walk upstairs. We opened mail, answered phones, and filled orders. It was definitely the most hectic time of our work year.

On one specific snowy evening, Mike, Matt, and I settled on the living room sofa, watching *Columbo* on television. As the TV detective snooped around an apartment building, we warmed ourselves by the flickering flames in our fireplace. Once the blaze died, I stirred the logs until the embers glowed an iridescent red. With dinner over and dishes finished, we sat cozy in the living room while a snowstorm moaned and raged outside. Unexpectedly there was loud pounding on the front door. Mike pushed himself off the couch and wandered to the entrance hallway.

"It's late for callers," he said. "I wonder what's up?"

Just as he opened the front door, a gust of wind blew in our neighbor along with a blast of snow. "Dean! What're you doing here? What's wrong?"

"We've slid at the corner," Dean answered. "Marsha's in the car but there's no way we can get out. We're trapped in the borrow pit."

Dean brushed snow from his shoulders, shook his wool cap, and rubbed his hands together as he stood in our stone-floored entrance. He wore a dark suit and obviously had not dressed for trudging through snow from the ditch along the roadside. Once Mike changed into padded overalls, Dean followed him to the garage. Grabbing his down jacket from a hook beside the back door, Mike added gloves and a wool cap.

"Here, take this," Mike said as he threw a jacket to Dean to put over his suit. "I have some extra gloves in the pickup."

The two of them climbed into Mike's truck and headed toward the ranch headquarters to pick up a tractor with a blade. I shut the fireplace glass doors and gripped Matt's hand along with a load of blankets. We ventured into the garage and donned heavy clothes, gloves, and boots. Once we scrambled into my car, I backed out of the garage. The wind picked up and snow blew against my Subaru, encasing the windshield with ice. I engaged the wipers and added heat. In a minute the wipers had knocked off the snow; and I put my car into second gear, slowly moving out our driveway, toward 4900. There at the corner sat an older car, half in and half out of a deep ditch. Its driving lights gleaming through the snow, but no one was in sight.

"Matt, you stay here until I know what's up."

I trudged through the snow and across the road to the trapped vehicle, bending into the wind and shielding my eyes from the blistering flurries. Just as I reached for the door handle, the face of a frightening clown flashed at the window. All black and red with orange hair sticking out from above its ears. I jumped backwards just as the window rolled down.

"Am I glad to see you," Marsha said. "It's freezing out here."

"Gads, Marsha, you scared me to death," I exclaimed. "Get in my car. I have the heater on and the guys will be here shortly. They went to get a tractor."

Once Marsha and I lumbered back through the snow and joined Matt in my Subaru, she wrapped herself in a blanket and explained the situation.

"I was giving a clown presentation for several youngsters at the Church. I kept on my makeup because Candy had a few gals at the house, and they wanted to see what I looked like."

"Did anyone stop to help?" I asked.

"One car slowed and then kept going," she answered.

"Can you imagine if someone came to the window, like I did, and you popped up as a clown?" I said laughing at her wild makeup. "He might have had a heart attack. Or shot you!"

Lights from the ranch's tractor glowed in the distance as it slowly advanced toward us. Mike drove with the blade down, slicing through small drifts along the road. As soon as they stopped, Mike jumped from the tractor, carrying a thick nylon rope. He crawled underneath the car's rear bumper, pushing snow to the side as he moved beneath the trunk. Using a flashlight to see, he attached the towline around the car's frame. As Mike squiggled backward, Dean stood on the snow-covered road, his suit and jacket collars turned up against the wind. Mike hooked the towrope to the tractor blade while Dean staggered through drifts to reach his car. Once inside, he put the gear shift in neutral and waved out the window, his thumb high in the air. Mike tightened the rope and put the tractor in reverse. Once it was taut, the tractor stalled. Then with a burst of energy, the tractor again engaged and pulled the car from the ditch. Yes! We raised our fists and cheered from the warmth inside my car.

"Call when you get home," I instructed. "We want to be sure you make it okay. We can get Mike's jacket later."

"Will do. And thanks loads," Marsha said as she opened the door and ran across the road to her car. She gave a final wave before disappearing inside.

Dean and Marsha pushed off first. Matt and I stared over the dashboard as they faded into the snowstorm. I followed close behind. Mike plowed the road corner and built a bank of snow in front of the borrow pit. Then he turned the tractor toward the ranch head-quarters. When Mike finally returned home almost an hour later, light from our house windows shined over the snow. I had already heard from Marsha and had a large cup of hot chocolate waiting for Mike. *Columbo* was long gone, and we were soon ready for bed.

Chapter Twenty-Three

Smoky in a Blizzard

"Don't worry. We'll be able to drive to Twin without any problem," Mike said as he checked the outside weather.

"Gosh. It looks terrible to me," I said as I grabbed my purse and walked Matt to Mike's pickup.

Later, during another dismal January afternoon, the sky had darkened to the color of purple plums, and snow began to fall. Soundlessly and steadily. For Mike this was typical of winter storms in Magic Valley. For me, it looked pretty intimidating. We drove through blowing flakes of snow to the Strand's house for Friday Night Bridge. Their two teenage daughters volunteered to babysit Matt while we played cards with eight other couples. It blew all evening while we drank wine, nibbled on appetizers, and played bridge. We sat cozy and warm in their house as the snow fell in unbelievable swaths, completely covering the outside.

On our way home a few hours later, ice had covered the landscape and power lines sagged. Then came the wind, a direct hit from Siberia. We forced our way through, what was by then, a fierce winter blizzard. In the hours we had been inside playing cards, a foot of snow had fallen. The wind whipped Mike's pickup, and streaks of snow blew across the windshield. We passed several stalled vehicles; indiscernible white mounds lined the sides of the road. The storm was much worse than we had expected.

Finally, we turned off Highway 30 to 4900, and rammed through two-foot drifts. I twisted in my seat to check on Matt in his rear car seat. Out the back window I saw snow billowing behind us like the wake of an ocean liner. From the front window, I glimpsed colorful Christmas lights protruding from snow-covered shrubs bordering the front of our house. When we plowed into our driveway, it was past midnight. Mike pulled into the garage, turned off the truck, and walked into the kitchen. I gathered Matt from his car seat and ushered him upstairs.

Wanting to check on Smoky, I changed in the garage from dress shoes to rubber irrigating boots, tucked in my blouse, and grabbed a hooded down jacket from the clothes rack by the back door. Once I opened the door, a strong gust struck my face and a bitter wind whipped the wool scarf from around my collar. I pulled it back to circle my neck and tightened the ends. Shielding my eyes with my right arm and bending into the gale-force wind, I trudged through the blinding snow toward our small barn. The wind howled as I moved away from the protection of our brick house. The halogen bulb high on the barn's gable had brightened the area near the barn. Snow fell in horizontal sheets, slicing through its yellow glow.

A large, dark mound lay on the icy ground underneath the diminishing light. As I came closer, I saw it was Smoky. He didn't move. I thought he was dead. As I stopped near his body, he raised his head a few inches, his nostrils extended. He kept his eyes closed. I reached over and grabbed his frozen forelock and ice-packed mane and tried to get him to stand. He was as heavy as a freight train.

"Come on, Smoky. Get up," I shouted over the wind. His rear legs bent backward at an awkward angle. He was on his side, his body stuck to the ice. The blowing snow thundered with such force, I could barely keep my balance on the slick ground. Leaving Smoky, I slowly pushed against the blizzard's pelting snow and returned to our back door.

"Mike. Come quick! Smoky's down. I can't get him up," I yelled from the kitchen.

"Damn. I'd just fallen asleep," he mumbled from our bed. "Okay. I'll get dressed."

I left the house and hiked back through the bitter cold to retrieve a halter and lead rope from the nearby shed. Mike changed into padded overalls, a heavy jacket, work gloves, and a wool cap. He telephoned Melvin once he was dressed.

"We need help. Smoky's down. Can you come?"

"I'll be right there," Melvin responded in a sleepy voice. "Judy'll make some mash. I'll bring blankets."

Mike pulled down the ear flaps on his wool cap and left the house. When he reached Smoky, I had already put on his nylon halter and attached a lead rope. The snow blew inside my upturned collar and froze around my neck. I shivered as we tried to get Smoky to stand. Looking through the glare of the barn light illuminating the wind-blasting storm, I saw headlights shining into our pasture. Melvin and his wife, Judy, had arrived. They held hands to balance themselves against the wind as they trudged toward us, bundled in heavy winter clothes, carrying horse blankets and hot mash. Just as they stepped on the ice, their feet came out from under them. In a second, they both fell and hit the ground, right on their butts with their feet sticking up in the air. Rolling over and crawling forward on their hands and knees and balancing the hot mash, they regained their footing and stood, inching their way nearer to the barn.

The blizzard continued unabated as the four of us pushed and pulled, encouraging Smoky to stand. The ground was too slick for him to get any traction. We took piles of straw from the barn and stuffed it under his frame, trying to separate his body from the freezing ground. We made a nest of straw around him so if he did rise, he had something besides ice to stand on. Mike took the lead rope I had attached to his halter and pulled, leaning his body backward against the rope. Judy and I pushed his rib cage, and Melvin used his weight to heave his hind quarters. Smoky seemed to have given up. He lay his head back on the ice and didn't move.

"Come on, Smoky. Get up. We can't do it without you," I shouted over the brutal wind. Mike let out a shrill whistle while Melvin and Judy made clicking noises, coaching him to stand.

"One. Two. Three," Mike hollered, a coat of ice forming on his mustache. We pushed and pulled and Smoky raised his head. He stuck his fore legs out in front and managed to get his hind feet under his hips. He stood in the storm and shook like a wet dog. We had been out in the roaring blizzard for almost an hour. Who knew how long Smoky had been there?

No way could we have gotten him upright if he had not tried to help himself. Judy and Melvin retrieved more straw and made a path to the barn. Smoky stumbled over the straw-covered ice while I held his halter and patted his shoulder. With his head hanging down, he struggled forward. Judy and I stood on one side of him, with Mike and Melvin securing him on the other. The wind hissed through the snow and continued to batter us. We edged him toward the shelter, holding on to both of his sides, guiding him through the blizzard.

With streams of breath coming from all of us, we used cotton rags to wipe his trembling body and legs. We covered him in blankets while he stood visibly shaking from the cold. Judy gave him a bowl of gruel: a warm mixture of rice, milk, and oats. With stiff joints and awkward movements, he walked from the barn's wide opening and into his box stall. I wacked my hands together, beating blood back into my frozen fingers as the storm continued to rage. Melvin and Judy cautiously made it back over the ice and returned to their pickup. Their truck faded in and out of the white swarm of blowing snow while I dug my cold hands deep into my pockets. Mike retraced his steps through the blinding blizzard to our back door, anxious to get inside out of the wind. I stayed to fill Smoky's stall with two feet of straw, creating a thick, forgiving cushion. As

I left the barn, I dropped the heavy canvas tarp covering the front opening and tied it in place. The heat lamp above the water trough would soon warm the area.

I trudged through the snow, pelted with ice crystals, and finally made it back inside our home. I was shaking as if I had been dunked in freezing water. I had not bothered to change my dress slacks that I had worn to the Strand's party earlier in the evening. Numbing cold erased all feelings in my legs and toes. I stood in our bedroom, my teeth chattering as I hugged myself and shivered uncontrollably. Peeling off my bitterly cold clothes, I pulled on one of Mike's flannel shirts and crawled under our thick bed covers. In between rumpled bedsheets, I moved toward Mike, trying to warm up.

"Christ! You're freezing," Mike exclaimed as he pushed away from me and moved to the far edge of the bed. Huddling by myself, I bunched the blankets around me and finally warmed enough to fall asleep.

The blizzard blew itself out during the night. The next morning, unusually sunny and bright, I called Dr. Monroe. He and his assistant, Ruth Sievers, drove that afternoon to Sky Ranch. Once he turned south on 4900, he saw the road had yet to be plowed. Being a long-time, big-animal veterinarian, he put his truck in four-wheel drive and pushed through the drifts. The back road looked like it had been covered by puffy quilts with streaks of blue ice interspersed between white snow mounds. From inside the barn, I saw out in the field frozen tumbleweeds resembling balls of glass shards sparking in the sun. Then I heard a truck pull up and squinted in the afternoon sun as Bob and Ruth climbed down from the vinyl front seats.

Inside the warm enclosure, the slight smell of manure permeated the air. Bob listened to Smoky's heart and ran his hands over his spine and down his legs. After giving him an antibiotic shot and some inflammatory medicine, he suggested I keep a blanket on him for another week.

"His thick winter coat will soon warm him up," he stated. "And he looks pretty good. But his back legs splayed on the ice," he continued. "He'll be sore for a bit. Thank goodness you checked on him last night. He's one lucky horse."

Chapter Twenty-Four

A Guest from Pennsylvania

"I can't believe how big they are," Susan exclaimed as she stood close to two ranch combines. "My gosh, the tops of the tires are over my head."

"That's my favorite machine," I told her. "At harvest time," I continued, "Matt and I ride with Mike during cocktail hour. Instead of wine and hors d'oeuvres, I bring cold sodas and cookies . . . enough to hold him 'til he gets home for dinner."

My college roommate, Susan Church, had joined me at Sky Ranch. She had never been on a ranch, and I was eager to show her Idaho's wide-open spaces and my adopted lifestyle. I drove her around the miles-long boundary of Sky Ranch and stopped at its corporate headquarters, showing the control center and their numerous farm machinery. From the main section, we moved to the parts room.

"Whenever anyone goes to town, they have to call the ranch to see if any parts are needed," I said. "After spending all day running my own errands as well as grocery shopping and picking up horse food, I hate calling to see if anything is needed."

We sauntered outside and I showed her a huge scale placed in the ground beside the building. Farm trucks drove over the cement slab to calculate the "before and after" weights during harvest. With these figures, the ranch could compute their crop income. Behind the headquarters Susan saw a dozen farm vehicles, red International combines and tractors and pale-yellow trucks. They were lined in rows like Revolutionary soldiers with an elevated gasoline container nearby.

"The farm has to be as self-contained as possible, including having constant gas for all its equipment," I explained. "Remember in '79 when we waited in long lines for gasoline? Well, the ranch never had to worry. The government knew people needed food, so they weren't going to let farmers go without fuel for their equipment."

From there, we tramped across the dusty, dirt road to one of the new potato cellars. I pointed to an older one still standing down the road. The main portion of the 1950 cellar was underground, hence, the name "cellar." Two-foot strands of grass grew upward from the sod roof, acting as insulation for everything inside. The yellow-tinged grass caught the morning rays in a golden glow, looking postcard perfect.

"It's used for only short periods," I said. "Any potatoes stored in the dirt cellar are sold as soon as possible."

The new potato cellar, made of vinyl, aluminum, foam, and concrete was air-conditioned and completely above ground. It stored potatoes for months. Consequently, Sky Ranch could sell its crop when potato prices were high instead of directly after harvest. Susan and I climbed the thirty-foot, metal ladder inside the cavernous building. At the top, we looked at a dark, empty cellar.

"It looks like a football field," Susan exclaimed.

"Not quite, but it is huge," I said. "In another month, this will be one busy place. A conveyer belt will move the potatoes from dump trucks into the cellars. Harvest time is fabulous with all the heavy equipment, trucks, and farm activity. It's like an insect colony with everything moving in all directions. I find it to be best time of the year."

We returned home for lunch, and afterwards, I introduced Susan to her temporary horse, King. Mike had borrowed Don's horse, a large reddish-brown gelding, for Susan to ride during the week. King emerged as the alpha horse and intimidated Smoky from the moment they met.

"Whenever you see a single wire around a corral," I explained to Susan at the edge of the pasture, "it's probably hot."

"What's that mean?" she asked.

"It's electrically charged. If an animal touches it, he'll be shocked. The horses know that and will stay away."

"Have they been shocked?"

"Yes. Definitely. But once they've been shocked a couple of times, we don't bother turning on the electricity. They won't get anywhere near it."

"How do I know whether it's hot or not?" Susan asked.

"All you have to do is touch it."

"But I don't want to get shocked!" Susan exclaimed.

"That's why you touch it with the back of your hand. If you're shocked, your hand grabs air," I explained "But if you touch it with the front of your fingers, you'll grab the wire."

"You certainly know a lot about farming," Susan said.

"Yup, I do now. But I was a complete novice when I first arrived," I said. "I'm still learning. It seems to be a never-ending quest."

After the electricity lesson, we rode Smoky and King to the gravel pit. It was a pleasant ride, seeing BLM prairies around us and high mountains in the distance. The immense landscape never changed as the isolated valley was so widespread. The cloudless, blue sky stretched from horizon to horizon. It was definitely big-sky country, not another vehicle crossed our paths as we meandered along the dirt road.

As usual, I wore a riding helmet, short boots, and half chaps over my straight-legged jeans. Susan sat on a Western saddle and wore jeans and sneakers. As we trotted down the dirt road in front of our house, Smoky shied at invisible shadows next to large boulders lining our path. This was a constant challenge for me as I never knew when he'd jump sideways, fearful of some unknown terror. We rode into the rocky maze of the gravel pit and through a jumble of rocks. When we exited the other side, Susan noted the extent of our remote valley, flat and vast without another tree or stream in the area. This was the barren land I had grown to love.

On our return, we put the horses into a slow canter. They knew they were going home and picked up the pace. Once inside our pasture, we dismounted and I unfastened Smoky's saddle. I snapped the stirrups up to the top and placed the saddle on the ground near the fence. Then I helped Susan remove the heavy, Western saddle from King. Holding their leather reins, we walked around the interior of the field, cooling off both horses. After we rubbed them down, we undid their bridles. In the barn the amber spiral of fly paper was completely studded with black dots. With two horses enjoying the enclosure, the flies had multiplied like gnats on a banana.

The following day we woke to puffy clouds as morning light slipped from the sky. Once breakfast was completed, I dropped Matt at Kay's house and took Susan to Twin Falls. Immediately upon entering the city, I drove through a carwash. A cloud of dust covered my car, not only from dirt and gravel roads, but also from fine earth powder seeping under our garage doors. Farm machines regularly shifted through nearby fields and with them came the ever-present dust.

After driving through the tree-lined downtown, we crossed the railroad tracks to park in front of the Livestock Commission Company. I ushered Susan to the rear of the building where we climbed a wooden staircase to platforms, protruding out over several holding pens. Each enclosure contained an assortment of animals: horses, hogs, sheep, goats, and cattle. After we left the holding pens, we entered the front door and heard the auctioneer calling out numbers from his microphone. Sitting in folding chairs on a raised row, we watched the proceedings. First, we saw sheep herded into the thirty-foot, circular space. The auctioneer, along with his assistants, stood on a platform above the sale area. A man with a red electric cattle prod, known as a hot shot, walked in the sawdust-covered enclosure. As sale prices were called out, he waved the wand to move the animals in a circle, then ushered them out a separate door. As that door closed, another opened. A Hereford cow and calf entered. Their brown and white markings were distinctive and seemed to bring a good price. The next animal to sell was a large black Angus bull. The price skyrocketed. Then a few more Herefords were shown. And so, it continued.

On our way back to the ranch, I headed east on Falls Avenue and turned north toward Shoshone Falls. We parked near a concession stand in front of the waterfall and walked down a dozen steps to a metal platform extending out over the canyon. The panoramic view of the magnificent surge and crash of the Snake River over the cliff was breathtaking. We listened to the sound of water smashing around gigantic boulders and plunging downward, dropping more than two hundred feet.

"It's higher than Niagara," I told Susan.

She took a few pictures as the mist from the waterfall rose into the air. A rainbow appeared in the sky and she took more photos. Finally, we climbed the metal stairs back to the parking lot. In an hour, we'd be home and ready to begin dinner.

After another morning of riding the following day, I talked Susan into walking with me to Larry Yamane's field, a property adjacent to Sky Ranch.

"Do you like beets?" I asked.

"Yes. I love them."

"Great. Our neighbor grows them. Let's walk over and dig some for dinner," I said.

Taking a shovel and walking with Susan and Matt a hundred yards to Larry's land, I dug into the soil. The dirt crumbled from the shovel and I smelled the fresh loam surrounding the beet.

"Wow!" Susan said. "They're gigantic."

"The sandy soil is perfect for beets and potatoes," I said as I stepped on the shovel's edge and circled the pale brown beet. It was the size of a small football. There was no need to dig any more. This one would be enough.

The beet was so large I had to boil it in my New England lobster pot. After an hour, I stuck a fork into its middle to see if it was ready to eat. Nope, it was still too hard. I cooked it for another hour then sliced it into cup-sized chunks and boiled it some more while Susan chopped tomatoes, celery, and green peppers for a salad. Matt followed us around the dining table, holding cloth napkins while we placed silverware and milk glasses in front of each chair.

Once Mike returned home, I baked some potatoes and broiled a steak, topping it with garlic butter. Salad, milk, bread, and butter were placed on the table. Dinner was ready. We sat in the dining room and Mike showed Susan how to pierce a potato. "You use a fork, not a knife. It helps with the taste and texture," he exclaimed. Then he told her about the ranch's preparations for harvest. "It starts soon, about the middle of August."

"What's this?" Mike asked as he poked at the beet on his plate.

"It's one of Larry Yamane's beets," I answered, proudly acknowledging my expertise of local farm crops.

"Are you kidding? Those are sugar beets," he grumbled as he pushed it to the side in disgust and rolled his eyes to the ceiling. "They're not garden vegetables."

Luckily, Susan had made a large salad and Mike had plenty of meat and potatoes to satisfy his hunger. Susan and I tried the beets, and yes, they were super sweet. You could barely taste the beet flavor as the sugar sweetness overwhelmed the vegetable. Idaho is one of the top producers of sugar beets, beets that are grown for human consumption as sugar, after they are processed. Not before.

There was another tourist attraction I wanted Susan to visit. It's known as "Balanced Rock," and is found in a deep canyon, south of Castleford. As a freak formation of nature, the rock is fifty-five-feet wide at its top and is held aloft by only a four-foot-wide connection at the bottom. It looks like a hand pointing. The section of the "wrist" seemed unbelievably tiny to support such a large boulder.

Parking with my car facing a rock wall, Susan and I walked up the trail to the rock. After taking numerous photos, we returned to the car. Once seated, I turned the key in the ignition. Nothing. I tried again. Still no reaction. We were stuck and sandwiched between two rock outcroppings. I walked to the road and flagged down a rare, passing trucker to see if he could help. He reached for his walkie-talkie and called AAA but he had no connection. We were too deep inside the canyon.

"I'll radio Triple A when I get on top," he said.

We waited almost an hour but no emergency vehicle arrived. Finally, two truckers stopped.

"Can you help us? My car won't start," I asked. "We should be able to push it backwards. Then, I can jumpstart it."

With two husky men beside me, we pushed the Subaru backwards and up the slight embankment. Susan sat in the driver's seat and turned the front wheels toward the road. From there it gained a good position to jumpstart the car. Susan and I switched places and I changed the gear from neutral to second. With my foot on the clutch, the car rolled further down into the canyon. As soon as the speed increased, I released the clutch and the engine came to life. It worked. I honked and we both waved to the friendly men as we proceeded down the hill. At the base, I finally had space to turn around. Once turned, I steered back up the winding road to the top of the canyon on our way back to Sky Ranch. Susan was impressed by the helpful truckers. I informed her that it was normal, and I agreed with her assessment. Living in rural Idaho meant people helped each other whenever possible.

On the last night of Susan's vacation, King decided to return to his home. He broke through the electric fence enclosing our pasture but didn't know the direction to his barn since he had been trailered to our place. Smoky followed in his footsteps. The two horses dashed up and down the hard-packed gravel road in front of our house, flying back and forth under heavy clouds. It was past eleven when we first discovered they had broken out. We had no street lights, and on that particular night, no moon illuminated the area. It was pitch black. The two horses were like fireflies, periodically flashing their bodies and then disappearing down the dark road. We could hear them coming. Galloping feet advancing toward us at full speed. I squinted at the sound and held Matt's hand, keeping my five-year-old close by. Our flashlight beams pierced the night as the noise came nearer.

A form appeared and King sped past me at a full gallop. I shielded Matt and Susan behind me. Neither horse wore a halter and both were frightened by the cluster of people standing near their pasture. Mike had called Terry and Melvin on the radio. They joined Susan, Matt, and me near the road in front of the house. Another ten minutes passed but the horses never slowed. The sounds of galloping hooves continued. The men tried roping them as they passed but the dark night made it impossible once they vanished from the glare of our flashlights. Smoky, the older of the two, began to slow. He trotted to the fence and stopped.

"Let me try," I said to the others. "He knows me and I have oats."

I walked slowly toward Smoky, using soothing tones and calmly talking. He stood still, knowing my voice and hearing oats rattling in a can. I held a lead rope behind my back, and shaking the can of oats in my other hand, I enticed him to me. He took another step forward and faced me. As he nudged his nose closer, smelling his favorite food, I slowly wrapped the nylon cord around his neck and cinched it tight. It was over.

Smoky followed me back to his pasture while King plodded behind. I ushered them into the small barn while Susan gathered some rags from a nearby container. We treated them to a handful of oats and a small flake of hay. After the men closed the fence gates and departed with Matt, Susan and I rubbed sweat from the horses. These heavy-breathing beasts were now as docile as could be. I touched Smoky's muzzle and he let out a soft whinny. He was home and happily content.

It was well after midnight when Susan and I entered the living room. We laughed about the latest adventure and recalled the numerous activities we had completed since her arrival. As Susan's last morning approached, light shown spectacularly on the eastern horizon. Once I pulled the car from the garage, we saw Smoky and King sprawled flat on the ground, soaking in the sun, their tails periodically flicking at nibbling flies.

The hours of galloping past our home the night before had exhausted them. By the time I returned from the airport, they had not moved. I paused, half turned, and looked back at them. Two old horses who had relished their night of freedom were appreciating the warm rays of the noon sun. It looked like two fat drunken sailors lying on a Caribbean beach. They didn't even lift their heads as I drove past their motionless bodies.

Chapter Twenty-Five

Young Matt Lost during Harvest

In autumn Matt and I often joined Mike on the combine while he harvested grain. This was our quiet time, no television and no telephone. Before I left the house, I'd call Mike on the radio to find out which field he was working.

On one particular Saturday, I took young Matt for a short hike to the arm of the center pivot behind our house. It was about the same time as we'd usually head to Mike's field. We were accompanied by Jack and followed by Buddy, our black cat. When we returned, I heard the telephone in the garage ringing. It was an Angler's business call. A few minutes later, I was off the phone, but neither Jack nor Matt were in sight. Buddy rubbed against my leg as I stared out toward the field we had just walked.

"Matt! Matt! Where are you?"

I called again and again. I yelled for Matt and went upstairs to check his room and my office. I ran outside and called again. I stood still, silently listening. There was no sound and no answer. The sun broke through bright clouds and bounced off our brass bell hanging beside the back door. I stepped to it and tugged the bell's rope. If Matt couldn't hear it, Jack certainly could. Our Lab would know to come back home and to drag Matt with him. But I heard and saw nothing.

I dashed to the barn and shed and raced around both of them. Nothing. Half a dozen professional hay mowers were cutting the field behind our house. With all the dust created by these giant machines, there was no way they would see a small child. I jumped in my car, honking and searching. Still no child and no dog. I returned home and radioed Mike.

"Mike. I can't find Matt! Clear." The radio cracked with static but my message went through. I didn't care if others heard my plight. I had panicked.

"I'll be right home. Clear," he answered.

He was in another field a few miles away. Terry radioed he'd shift positions and take Mike's combine. Once the change was made, Mike ran to his pickup and shot back to our home. Just as he passed the ranch headquarters, he saw our little tyke singing and skipping toward the main building. Matt bounded down the dirt road with our black dog wagging his tail and prancing right beside him. It was a beautiful day, and Matt was going to see his Dad.

"I think your Mom wants you," Mike called out his truck's window as he came to a stop. He walked to the duo and leaned over to address Matt. "And I don't think she's too happy."

He scooped Matt up and placed him in the front passenger seat. He then walked around the rear of the pickup and lowered the tailgate for Jack to jump in. They were almost a mile from our house. By the time they arrived home, I was terrified. I knew what had happened. Matt was not going to wait for me to get off the phone. He had decided to walk to the field and ride the combine alongside his father without me.

I reached out and took him in my arms, hugging him as I brought him inside.

"You're never to walk by yourself," I scolded. "With all the trucks around, it can be very dangerous."

"But Mom," he responded as he looked up with bright, innocent eyes. "I wasn't alone. I had Jack."

* * *

During harvest, everyone worked late—well into the night, until they couldn't see to complete a straight line through the fields. Fine powder from the fields crowded the air and overwhelmed the land. Streaks of red and yellow filled the horizon, and as the day turned toward evening, the sky changed to a reddish gold. A gigantic, orange moon appeared, hence, the significance of a "harvest" moon.

"Hey, Mike, which field are you in? Clear," I radioed from our kitchen around five in the late afternoon.

"Go three miles past Don's house and turn right. You'll see me in the field on the left, one mile up. Clear."

"Okay. I'll be there in ten minutes. Clear."

As I drove over ranch roads, my car became enveloped in a plume of dust kicked up by a large truck in front of me. Farmer's called its trailer a "belly dump" because its tank bed did not tilt backward. Instead, the V-shaped bottom opened and funneled grain from the tank into a container. From there, the crop shifted to a metal tube, transporting grain to the top of a storage tower. These silos peeked out from the landscape throughout Magic Valley.

After letting the belly dump move ahead, I parked at the end of Mike's field and waited for him. Once he turned the combine to start a new row, I trudged over the pulverized soil, bringing enough snacks to hold him until he came home for dinner. Mike opened the combine's door, and I grabbed a large bar and hoisted myself up the metal steps. Besides food, I also brought the day's mail. I sat on the small metal bench next to Mike as he drove the giant, red machine through the barley field. The harvested crop was destined for either Anheuser-Busch or Coors Brewing Company in Burley.

"Where's Matt?" Mike asked.

"He's at Kay's."

"What's in the mail?" he asked as he opened a Pepsi can.

"Nothing much. A few bills and a letter from Ron and Maggie." They were the couple who had flown to the ranch in their Cessna four-seater plane the year before. Ron had landed on the dirt road in front of our house and then had helped Mike harvest the last of the ranch crops. He had originally farmed a Midwest homestead and knew how to operate large equipment. He and Maggie owned Bristol Bay Lodge in Alaska and we often fished and photographed at their place during slow times at Sky Ranch. In the winter, they lived in Washington. Because Ron had assisted Mike the previous year, the four of us were able to travel to Central America on the dates we had originally planned.

"What's going on with them?"

"They want us to come for Thanksgiving," I said as I read the note. "We have nothing planned. Want to drive to Washington?"

"Sure. Maybe they can join us for another fishing trip in Belize," he said.

That sounded wonderful. I smiled at the suggestion and began to feel romantic. While Mike stared at the row in front of us, I leaned over and kissed him. There were three other combines in the field, stacked in rows behind us. The four combines crossed the field in a large swath, not beside each other, but arranged in separate rows, one behind the other.

"What's going on?" Mike asked as he smiled and turned toward me.

"Ah, nothing. Just thought you'd like a little break from all your hard work," I said as I slid my hand between his thighs.

"Wow. Now, what're you up to?" he asked.

"You just concentrate on the rows ahead. I'll do the rest," I replied.

He tossed the Pepsi can to the floor and grabbed the steering wheel with both hands. I removed my jeans and climbed onto his lap with my back to the field, straddling him and burying my face into his neck. He tilted his head and looked past me, trying to keep the combine moving in a straight line. How he kept harvesting, I'll never know. When I later climbed down the metal stairs, Mike wore a wide grin and waved his hand out the cab door.

"Come back real soon," he called, laughing as he closed the door and turned the combine into the next row.

Chapter Twenty-Six

Bart, the Gardener

Two churches provided spiritual faith in Murtaugh: The Church of Jesus Christ of Latter-Day Saints, or Mormons, and the Methodist Church. Since I knew nothing about Mormons when I moved to Idaho, I chose to attend the Methodist Church. The minister, Dale Metzger, telephoned me one day and asked if I would help a young Christian man who had had a bad turn of luck.

"Bart's a good guy," Dale said. "He got caught stealing food for his family. He's out of prison and needs a job. Can you help?"

"If he can do yard work, yes, we'll hire him," I said. "He can come over Saturday and work for a few hours. We'll see how it goes."

A few days later on a chilly early morning, Bart stood before me. He was tall and thin and wore new blue jeans and a pressed, white cotton shirt. I thought he looked overly dressed for pulling weeds, but I was pleased he had arrived on time. We chatted while I showed him which plants to remove and which shrubs to trim.

"My wife washes our clothes in the bathtub," he said. "We don't have much."

"I'm glad to help. And I'm glad you can help me with gardening," I said as I stood ankle deep in weeds.

"Thanks for the job," he said. "It's been tough since I got out of jail."

Mike had left earlier in the day and would not be back until lunch. Matt was at Kay's house, and I was home by myself. When talking with Bart, I noticed he had no teeth. His thin lips surrounded a dark gummy hole, and he talked with an obvious lisp.

"Did you have an infection?" I asked.

"All my teeth were bad," he said. "They removed them while I was in jail."

He bent to the earth and began to dig weeds and stir the soil, just as I had instructed. Having Bart relieve me of such tiresome gardening, I walked away from his strong cologne and entered the kitchen. In a few minutes Mike's lunch was made, and I returned to Bart, bringing him an extra sandwich and a soda. We took a break and sat on our cedar chairs on the patio, facing each other.

"Do you believe in Jesus Christ?" he asked. "Have you accepted the Lord?"

"Yes, I have," I said. "When I was in high school."

After lunch, the subject of Christianity continued as we dug weeds side by side. When Mike arrived home, I introduced the two men. Once he ate, Mike returned to the farm. Bart and I labored uninterrupted for the rest of the day. The garden patches around the front of our brick home were almost weed free. Our red, thorny barberry bushes had become unruly, and I asked Bart to trim them as one of his chores the following day. Then I pointed out some fruit trees that needed thinning.

"You'll have to work by yourself the next few days," I said. "I have chores to do inside."

Two days later, I brought out sandwiches, and we again sat out on the patio. He told me he didn't know where we lived when he first got the job, so he had driven down 4900 and parked in front of our house. He remained there and surveyed our property. His comment didn't bother me until he added, "I do that often. I like to watch you at night to see what you're doing."

That sounded creepy. I was not at all comfortable with the direction of his conversation. Leaving him to finish trimming, I moved to the back of the house and started planting flowers. Before long he came and stood behind me. I knelt in front of him, pushing the last of the zinnia plants deep into the soil. He persisted in reciting Christian versus and asked if I agreed with them. He must not have liked one of my answers because he took his trowel and slammed it back and forth into the open palm of his left hand. His voice alarmed me as he challenged my answer, standing close behind my kneeling body. The noise of the smashing trowel frightened me and an ice-cold shiver crawled down my spine. My tongue felt dry as I swiveled my head, still on my knees, my back to his legs. I bent one knee and stood to face him.

"Let's get back to work," I said as sternly as possible. "We only have this little area here beside the patio. Then we're finished."

Mike had been delayed and had not yet arrived home for dinner. I was frightened being alone with Bart. When Mike drove into the garage, I fled to the security of the house and hurried into the kitchen.

"Please tell Bart we don't need him tomorrow," I said. "Tell him we've finished the gardening, and I don't have any more projects."

I didn't want to confront the man myself. Once I told Mike about the incident, we started deadbolting the back door. The front door was constantly locked, but the back door, the one we used all the time, had always been left unlocked. Before long, we forgot about the incident with Bart and we went back to having an unlocked back door.

Two months later, the telephone rang. "Is this Bobbi?"

"Yes. May I help you?" I asked.

"Yes, you can. You can please me." The voice said slowly. "I want to lick your . . . ."

"Bart. Don't ever call me again!" I yelled into the receiver. "You can't fool me. I know your voice," I said as I slammed the phone into the cradle. I did know his voice. With all his missing teeth, I could easily recognize the lisp in his words. I informed the Methodist minister and then radioed Mike.

"I'm going to Twin. Need anything? Clear."

"No," he replied. "When will you be back? Clear."

"I have just one errand to run. See you at supper. Clear."

To protect our son who now slept in a bunk bed on the second-floor, quite a distance from our master suite, it was essential to have something to detect if anyone entered his room. I drove like a woman possessed and purchased a baby monitor at Sears. At dinner that evening I told Mike what had happened, and we installed the monitor in Matt's room with the receiver in our bedroom. We began locking the back door but I never felt completely safe after that incident. We lived so far away from our neighbors, and Mike could be gone on the ranch for hours at a time.

Chapter Twenty-Seven

Kindergarten Party

"Hurry," Matt said as he stood with an olive-green pack on his back. "Take it before the bus comes." At the start of his kindergarten classes, I told Matt I wanted to take his picture for the family album.

"Wow! Look at that," I said, pointing to our road out front.

On the dirt road bordering our house, four riders herded a large flock of sheep from their summer grazing land in the Albion Mountains to their winter pasture a few miles away. A half-dozen excited dogs ran back and forth, nipping at the sea of creamy-white bodies, insisting the band move forward. As they passed, they kicked up tiny dust devils, and we smelled the sweet fragrance of sage mixed in with the endless droppings of sheep. The bell of the lead ewe mingled with the noise of baaing sheep, the cracking of whips, and the whistling of cowboys. The sight of moving sheep in the background made for a perfect Western setting to Matt's first day at school . . . nothing like my mother's kindergarten picture of me walking hand-in-hand with friends down a paved Connecticut road.

Several minutes later, the yellow bus turned into our circular driveway and arrived near the front door. Matt marched off like a soldier intent on a mission and climbed the bus steps. He waved to Mike and me as he walked down the aisle and sat in one of the black vinyl seats near his friends, JR and Little Mario.

* * *

In May of the next year, I invited Matt's kindergarten class to our house to celebrate his birthday. I had decorated several card tables with pictures of dinosaurs, palm trees, and flying birds. Before they sat for cake and ice cream, we played pin-the-tail-on-the-dinosaur. Then I took a piñata, shaped like a small caveman, and hung it from our front hall balcony. It looked like a real person, hanging from our second-floor railing. Even more disconcerting was that it was the same size as most of the kindergarteners.

"One, two, three, four," I chanted as I touched each child's head when they gathered in the foyer. I pointed to the fourth child and the rest lined up in a row behind him. One of the mothers blindfolded the boy and the youngster swung at the piñata with Matt's aluminum baseball bat. He missed but continued swinging. He almost hit another child with his wild swipes. I quickly moved the line of children to the staircase overlooking the stone entrance. Now they could watch the action from a safe distance. What an uncomfortable scene of children attacking that little man wearing only a loin cloth, but the youngsters and their mothers laughed hilariously. Finally, one of the boys smashed the caveman and it broke into a mountain of pieces. The children ran to the shattered piñata and tore the caveman apart. Nothing. Then they turned and looked at me.

"Where's the candy?" the children yelled.

"What candy?" I asked the mothers.

"Candy?" they said in unison. "You're supposed to fill the piñata with candy!"

Coming from Connecticut, I had never experienced a piñata party. I didn't even know what a piñata was. When I purchased the paper-mache object, no one at the store told me to fill the object with candy. Although there was obvious disappointment, the children returned to the card tables, sang songs, and opened goodie bags. They each received a small sack filled with dinosaur erasers, candy, and pencils. Not all was lost.

* * *

A week after the birthday party, Mike planned an overnight business trip. Because it would take almost an hour for Burley or Twin Falls police to reach our house in case of an emergency, I often felt vulnerable whenever Mike was out of town, especially after the incident with Bart. On this occasion, I told Matt he could sleep downstairs in our bed.

"Let's play a game," I said. "We're going to tie ropes across the hallway. That way no bad guys can catch us in the middle of the night."

Matt and I wrapped a thick cord from the railing surrounding the pool-table area to heavy, upholstered chairs in the living room. Back and forth we wove the ropes. Matt crawled underneath to assure no one would be able to get below them either.

"This is really fun, Mom," he said.

In the master bedroom, I took an old ladderback chair and wedged it under the door knob. Instantly, I became comfortable. There was no way anyone could get to us. Or so I thought. Matt fell asleep beside me while I read a book. Finally, I tired and closed my eyes. About two in the morning, I heard an unusual sound coming from the interior of our house. I sat straight up in bed. Crawling from under the covers, I reached below the bed and pulled out my shotgun. I fiddled with the slide, opened it, and inserted a bullet. As Matt slept, I approached the bedroom door, placed the gun on the carpet, and silently removed the chair from under the knob. Inching the door open, I listened intently. Then I heard the sound of footsteps in the kitchen. I stood motionless, staring down the dark hallway.

"Who's there?" I shouted. "Come out now or I'll shoot. My gun is loaded."

Matt crept out of bed and approached me, pulling on my flannel night shirt. "What's happening, Mom?" he whispered as he looked at me.

"Stay back," I commanded. Matt moved to the side of the room and stared at me with sleepy eyes turning into giant circles.

Then I heard the sound again. I was terrified. I pressed my right check hard against the stock and pointed my barrel over the ropes and down the hallway, directly into the kitchen. I braced the end of the stock to my shoulder and prepared to shoot. Matt covered his ears. Just then I heard the sound of falling ice. I paused, turned on the hall lights, and looked toward the kitchen. No bad guys in sight. Thank goodness I hadn't shot. I would have blasted our refrigerator, and Mike would have really been pissed.

Chapter Twenty-Eight

Winter, Spring, and Summer Sports

When winter snows arrived later that year, Mike had stopped smoking, but he struggled to break the awful addiction. He wanted to stop, first for me, and then after Matt was born. But it took five more years to finally complete the transition. He paced the kitchen floor and chewed nicotine gum nonstop.

"Let's take the kids sledding," I suggested, trying to take his mind off smoking.

"Okay. Call them and let's meet here in a half hour," Mike said.

We stood outside and gathered the neighborhood children for sleigh rides over our ranch roads and nearby pastures. Mike took a long rope and fastened each sliding vehicle to the one in front: wooden sleds, toboggans, cookie sheets, and plastic slides. Anything that could skim over the icy surface was attached to the bumper of Mike's pickup. The kids piled onto their sliding objects, and Mike climbed into his truck. I sat upright on the first sled, facing the pickup with my knees bent and my hands holding onto its sides. Once I gave Mike the okay, we were off. With shouts and cheers, and with Jack running and barking beside us, we soared across the snow-covered fields.

When we abruptly came to a patch of ground without any snow, Mike made a quick turn. He didn't want the sleds to travel across dirt. The line of kids zipped in a circle and slammed into the side of a snowbank. Everyone stayed on and yelled for more. They thought it was great. I thought I was too old to play "crack the whip."

We continued at a moderate speed until I gave up my sled and sat in the pickup with Mike. As I watched out the back window, the kids yelled, "Go faster. Go faster." Mike gunned the engine and they were off again. Jack couldn't keep up while we raced across the fields. Before long, we arrived back at our house. The children trooped inside, rosy-cheeked and exuberant. They dropped their outer clothes at our back door and came into the house for popcorn and hot chocolate. Mike undid the sleds, closed the garage door, and recoiled the rope. Jack came through the doggy door with his tongue hanging out. Once he entered the living room, he slumped on the carpet next to the couch and fell fast asleep. It was soon time for supper. I drove the children home and returned to make a spaghetti dinner, this time without apricots.

That night, miles from civilization, I stood on our back steps and marveled at the sky. I had never experienced such beauty growing up, having always lived close to cities. Any night without clouds or a moon at the ranch meant the sky was filled with a lacework of stars, so abundant as to almost saturate the heavens. As I studied the sky above, I was struck by the silence of the night. No farm machines. No cars or trucks. Because so few birds or animals existed in that arid land, the lack of noise caught my attention. A chill came through the air, a breeze blew through our planted trees, and then nothing. The sound of silence.

* * *

In early spring we took to the slopes whenever enough snow still covered the ground. We wore the latest ski fashions and piled into

Mike's pickup, our long skis fastened to the top of his truck. A cloud of blond hair circled Christa's face as she climbed in, joining her Dad, Matt, and me on the ride to Pomerelle. We wore colorful wind breakers and black powder pants. The mountain resort boasted a temperature of fifty degrees, a warm welcome for a fantastic day of spring skiing. Living close to Pomerelle meant we could be on the chairlift in just forty minutes after leaving our garage. Mike took back roads as he wound his way to the lodge, passing barren boughs on leafless trees, shimmering with layers of icy crystals. Except on weekends, we never had to wait in line.

Matt started skiing at two but that only meant getting into winter clothes and walking around our snow-covered lawn in tiny ski boots attached to tiny skis. By the time he was five, we began taking him to Pomerelle. I tried teaching Matt by placing him between my legs and holding him up by his elbows. The two of us slowly skied down the beginner's run, a gentle slope close to the lodge. After an hour, my back felt terrible. Christa skied over to us, showing off her latest repertoire of innovative twists and stops. She looked at us and our strange stance.

"You know Matt's feet aren't on the ground, don't you?" she said as she laughed.

I looked down and, sure enough, as I had held his body, Matt had lifted his feet from the snow. In my bent-over position, I had been carrying this clever tyke down the hillside. No wonder he skied so skillfully and my back ached.

"That's it! You're going to ski school," I declared as I steered him toward a group of youngsters at the bottom of the slope. While the instructor taught him the basics, I enjoyed the next two hours of unlimited skiing. I felt exhilarated by the corn snow, small pellets of hard snow, along with unusually warm spring weather. Skiing in the West was so much better than skiing on icy slopes, the snow conditions I had experienced growing up in the East.

"He's ready now," the instructor said.

"What? Two hours and he's ready?"

"No. Not by himself. Here, take a rope and tie it around his waist. You'll snow plow behind him. He'll be fine."

And that's exactly what we did. With the tips of my skis together and the ends of my skis far apart, I snow plowed behind Matt. He built his confidence and skill, and I renewed my back muscles. Within a year, he was jumping moguls and racing with friends from the top of the mountain to the lodge at the bottom. Having Pomerelle so close to Sky Ranch made for easy access to moderate-level skiing. Having Sun Valley about two hours away made for tough skiing as it had a thirty-four-hundred-foot vertical drop, over a hundred trails, and eighteen chairlifts. Because we lived in the Rockies, that meant we had plenty of outdoor activities all year long.

* * *

When the following spring blossomed into summer, I enrolled Matt in tee-ball, a ten-week program for young children, ages five through eight. Getting to know Murtaugh people was not easy as Sky Ranch was located so far from town. As most of his classmates lived in town, playing tee-ball gave him a chance to make more friends and for Mike and me to meet socially with their parents. We assembled at the municipal park, nestled within a grouping of tall, elm trees. Mike and I carried aluminum folding chairs from his pickup and placed them in a line between home and first base. The audience cheered and clapped and filled the park with laughter. When the whistle blew, all the children, wearing shorts and T-shirts, ran onto the field. They tried to find their designated spots and ran like bumbling bees as they raced from one area to another. Frustrated coaches shouted instructions and eventually both teams were assembled in their correct positions.

Because most action took place around the infield, the three outfielders looked bored and eventually sat down. One picked dandelions, another lay on his stomach with his head resting on his hands, and the third player sat cross-legged with his back to the game. Parents yelled at their children and coaches encouraged the outfielders to stand and pay attention. When Matt came to bat, he hit a ground ball from the stationary tee and ran to first base. The next batter hit the ball and Matt ran to second. We rose and cheered as Matt began to run to third. The opposing players kept fumbling the ball. Matt had a chance to run for home. Mike stood in front of his chair and cupped his hands, yelling for Matt to keep running.

"Go Home! Go Home!" the parents of our team shouted from the stands.

Matt raced around third base and ran straight to his teammates who were sitting on a side bench. He thought that was "home." He never touched home plate. His coach pushed him back to the playing field and pointed to home plate. Matt touched the plate well ahead of the roving baseball. The stocky volunteer umpire kindly ignored the mistake, and Matt's team scored another run. With happy smiles we gathered our chairs and returned to the pickup. Matt's team had won the game.

After tee-ball season, we instigated a Fourth of July celebration, and invited neighbors and family to join us at our house. Melvin, Terry, and Mike fetched a slew of fireworks they had recently purchased from a traveling vendor. For years, this salesman had peddled colorful explosives he had bought in Wyoming, selling them in Idaho where they were illegal.

As guests arrived, they carried food and folding chairs to our backyard. Mike, Christa, Matt, and I had set up a half dozen tables and placed them on the cement pad behind the garage. The visitors deposited their food, and I brought out a large bowl of punch. Mike's son, Blaine, lugged a cooler filled with beer and ice to a side table. Chairs were assembled in a semi-circle out on our lawn, facing Smoky's pasture with a grain field beyond and the rising foothills as a backdrop. While the women supervised the food, Terry, Blaine, and Melvin carted a wooden platform to the edge of our lawn. This would be the stage. The men assembled fireworks to be ignited in a specific order: from small to large and back to small again. At the end, they would have the grand finale.

When the sun set and the sky darkened, we took to our chairs as if a bell had rung. It was time to start the show. The first explosives went as planned, sky high with white and blue sparks falling to the ground. The bright colors illuminated our raised faces and brought cheers from the crowd. For the rest of the evening, we clapped and shouted our approval while children ran in circles and carried sparklers high above their heads. They screamed in delight whenever flashing fireworks splashed the sky with colorful glitter. During the finale, one errant rocket came screaming toward us, flying beneath our chairs. I jumped forward and immediately stamped it out.

The next one in this lively group of rockets indicated the men had lost all control. Explosives shot into the edge of the grain field. A small fire started. Everyone vaulted from their seats and ran into the field with Christa and Blaine leading the way.

"Get up!" Mike commanded everyone. "We've got to get this out!"

The group stamped on the sparks, and I ran for shovels and brooms. All thirty of us were in the field, children and adults stomping and shoveling. It took time as the field was exceptionally dry. Finally, the fire was out and the nearby grain field was saved. We were exhausted. The show was over, and it was time for everyone to leave. They gathered their leftovers, carried folding chairs, and headed to their cars. No way would they forget that unusual Fourth of July celebration.

Chapter Twenty-Nine

Polly the Pig

In September after attending an airline reunion in California, I flew on a Sunday afternoon from San Francisco to Twin Falls with a plane change in Utah. When I checked the departure board in Salt Lake City, I saw my Delta flight had been delayed five hours.

"My, gosh. I can drive home in four," I thought.

While waiting, I read the *Salt Lake Tribune*. Out of curiosity, I opened the classified section and ran my fingers down the automobile column. A two-door Cadillac was listed at a reasonable price. At a nearby payphone, I called the ad's number. The owner answered and I told him I was at the airport on my way to Twin Falls and I was interested in his car. He said he was a physician who traded his car every two years.

"I'll drive to the airport and show you the car," he stated. "You can test it. I'm sure you'll like it."

"Great. I'm wearing black slacks and a pink blouse. You won't miss me. I'll be standing in front of the Delta baggage claim." Not wanting to annoy Mike, I didn't inform him. Since I used the Angler's funds, I thought I'd just surprise him if the owner and I came to a decision.

When I left the area to rendezvous with the car owner, I stopped at the information booth, located at the front of the airport. The grey-haired woman behind the counter smiled as I approached.

"May I help you?"

"Yes, you can," I nodded. "I'm leaving in a few minutes to test drive a car. If I'm not back in an hour, will you please call the police?"

"Really? You don't mean that, do you?" she exclaimed. "I've never had such a request. Gosh. That scares me."

"Don't worry," I said. "Here's the telephone number of the car's owner."

As I left the airport, I turned and glanced back at her. She looked like she was going to burst into tears. She was on the phone, probably asking a supervisor what she should do.

I hope this won't take more than an hour, I thought. *Or I'll be in big trouble.*

The doctor parked in front of the baggage-claim area and waved to me as he climbed out of his car. He sported gray slacks and a navy blazer. We introduced ourselves and I walked around the vehicle. It looked brand-new. He let me drive on several nearby roads and the Cadillac functioned flawlessly.

"Well, what do you think?" he asked. "Is this something you'd like?"

"Yes, I do like it," I said. "But let's talk price."

Once we decided on cost, he suggested we drive to his home since he and his wife lived close by. On his dining table in his one-story house, we wrote an agreement. The area had white carpeting and sparkled with cleanliness. His car was just as spotless. Our handwritten contract stated my bank would wire money from my Angler's account to his bank on Monday, the next day.

He gave me the keys to his luxurious car and I drove back to the airport, staring at the hood ornament and thinking what a great deal I had just made. Yet no money had changed hands. He trusted me enough to give me his expensive automobile with nothing more than a handshake and a piece of paper. This business transaction was another reason why I loved the Rockies. Life was good and honesty was paramount. I immediately reported to the woman at the information booth. She leaned across the counter and grabbed both my hands.

"I'm so relieved. I've been terribly worried," she said. "You shouldn't do these things. You could get into trouble. You're lucky the owner was decent."

I nodded, thanked her for her concern, and walked from the airport. My new car waited outside. I'd be home in four hours. This action was typical of the way I lived my life. I had seized another opportunity, another spontaneous decision.

After the purchase of this fancy car, my time lapsed into normal ranch routines. I worked weekdays upstairs in the Angler's office and on weekends I joined Mike and Matt for fishing and camping trips. When Halloween approached, Mike took off for a week of fishing on the Clearwater River in Northern Idaho. While he was gone, I bundled eight-year-old Matt and Kellen Nebeker into the car and gathered his two older brothers. The five of us were going roller skating at Skateland in Twin Falls. After a couple of hours, I collected the four boys and watched them race back to the car. Taking the Foothills Road to Murtaugh, I pulled into the Nebeker's driveway.

"Would you like to see our baby pigs?" Kellen asked. "They're just born."

"Sure," I said. "I've never seen a baby pig."

We walked toward the barn at the rear of the property. The boys' father, Bill Nebeker, accompanied us into the weathered building. I was surprised to see the sow had been placed in a V-shaped, metal stockade. An enormous pink hog lay within the enclosure while a bunch of little piglets ran between her legs, each trying to get a good hold on one of her teats. The wide section of the "V" was open at the top, the bottom section was tighter but allowed the sow to lie down while nursing.

"Why's she so locked up?" I asked Bill, who stood next to me.

"That's so she doesn't crush her babies," he answered. "She's huge, so if she went down quickly, she could smash anything below her. This way the piglets can nurse and she can't accidentally hurt them."

"Would you sell one?" I asked impulsively. I knew nothing about pigs and neither did Mike. The only hog farmer living nearby was Bill and he only had a few. I must have thought this little pig would be like Wilbur in *Charlotte's Web*. So, I responded as I did with most things in my life—I had an opportunity and was willing to take a chance—just like the Cadillac purchase in Salt Lake City. Besides, those piglets were darn cute!

"Sure. At this stage, they only weigh eleven pounds," he answered. "How about fifteen dollars?"

"Great. You only have one in black. May I have it?"

"Of course," he said as he reached into the pen.

While I went back to my car to retrieve some money, Bill deposited the squalling bundle into a cardboard box and carried it to my car. It was as little as a month-old Labrador puppy. He placed the open box in the front seat and gave me a small bag of food.

"This will hold you over until you can get to town," he said. "She's only three weeks old but she'll be fine without nursing. This kibble is specially made for piglets."

Matt pushed the driver's seat forward and climbed into the backseat. He stood, craning over the front passenger seat, staring at the squirming object in the square box.

"She's a Yorkshire," Bill added as I pushed two buttons on my armrest and lowered both front windows.

Once we cinched our seatbelts and waved goodbye, I backed out of their driveway. We had a few miles to go before we reached the ranch. The inquisitive piglet peaked from the box and stared out the open window. How strange to see this tiny, black pig sitting in the front seat of a Cadillac, checking out the landscape, its little ears flopping in the wind. As a child, I grew up with numerous pets: dogs, cats, parakeets, canaries, turtles, and fish. Having a piglet meant having another pet for Matt and me to enjoy. Mike wasn't into pets. He thought, as did most farmers, that animals were useful only for food or as work animals.

On the way home, we talked about different names. We decided on Polly. After I carried her into the house, Matt made a safe spot for her in the laundry room near the backdoor. He brought in newspapers, layered them on the vinyl floor, and placed old towels at the far end. I added two bowls, one for food and another for water. In our garage, I found a large piece of cardboard and cut it to hinge into the pocket door's gap. We could look over the two-foot-high cardboard gate and check Polly's status. I gently lowered her onto the laundry room floor and the two of us watched her scramble from us, squealing in the process.

Once she seemed settled, I looked up Yorkshire pigs in my *World Book Encyclopedia*. "They are white or pink with upright ears," the book stated. Her mother and siblings were as described. Polly was black with drooping ears. She must have been an anomaly as she certainly didn't look like any other pigs at the Nebeker farm. The encyclopedia also said Yorkshires can live up to ten years. Obviously, that would be rare as most pigs are slaughtered between four months and a year. That is, unless they are used for breeding. Matt and I quickly learned two things about pigs: one, they are slobs, and two, their manure stinks. When we checked on Polly ten minutes later, she had shredded the newspapers, turned over the water bowl, and sprayed pig food everywhere. Her home looked like a literal pig pen.

In another hour of a whirlwind mess in the laundry room, I opened both the kitchen and back doors and let her outside to do her business. Polly went around our arbor vitae hedge, into my flower bed of orange gladiolas and purple pansies. She accomplished her mission, but from then on, all hell broke loose. She ran onto the lawn and seemed impossible to catch. I had no idea a tiny piglet could run at such a speed. With both Matt and me chasing her for twenty minutes, she finally tired enough for us to trap her. She was only three weeks old and it took all that time for us to corral her. I got an old dog collar, put it around her neck, and added a leash. No way would we ever let Polly loose again. We were absolutely beat!

The next day, I once again attached the leash to her collar and opened both the kitchen and back doors. I followed her as she went right to the same spot, turned around, and did her business. She then walked back into the laundry room. She had been house trained in one day.

Life went on as usual for another few days until Mike called to tell me about catching some steelhead on the Clearwater. Because he was in such a good mood, I thought I'd bring up the subject of my latest pet. We already had a black Labrador, a black cat, and a black horse. Why not a black pig? Obviously, this was not what he wanted to hear.

"That pig better be out of the house before I get home!" Mike roared into the phone. The line went dead.

"You're so mean." I whispered to empty air and slowly replaced the receiver.

"That didn't go well," I said to myself. I called Terry on the house phone and explained the awkward situation about my new pet.

"Can you create a partition inside the barn for the pig?" I asked. "Mike is coming home tomorrow, and I have to get her out of the house."

"Melvin and I will come over tomorrow afternoon. I'm super busy right now," he answered.

"Thanks. You've saved my hide. Again!"

The guys didn't arrive until late in the day and didn't have the right supplies to finish the job. Mike would be home that night and I was panic stricken.

"Is that pig out of the house?" Mike asked when he telephoned from Boise, three hours away. I explained that Polly was still in the laundry room because Terry and Melvin couldn't complete the job as soon as they had expected.

"They promised me it would be finished tomorrow morning," I said. "Can you get a hotel room tonight?"

"No. I'm coming home. And I'm dead tired," Mike answered sharply.

I knew he would be exhausted as it was at least a ten-hour drive from the Clearwater River to Sky Ranch. When he walked in the back door, I stood in front of the kitchen counter, mixing salad ingredients. I turned and faced Mike. His eyes were hard and cold. I braced myself against the force of his anger. He frowned and walked straight into the kitchen. He never looked into the laundry room, not even to glance. Not even out of curiosity. I gave him a hug and again explained the situation. He didn't care. He just wanted something to eat and to rest on the living room couch. It was early evening and he was totally drained. While Mike brought in his suitcase, I placed his dinner on the kitchen table along with a pitcher of milk.

"It's Halloween so you'll have the night to yourself," I stated after he had eaten his dinner. "I'm taking Matt and a few of his friends Trick or Treating. We'll be back in a couple of hours."

"Good. No one to bother me," he said as he headed to the living room. Alone, he would be able to doze in the quiet of an empty house.

After a perfunctory hug, I left to gather Matt and some neighborhood children: Sarah, JR, and Little Mario. They huddled in the back seat and Matt sat up front. We covered at least forty miles to reach the homes and ranches around the south side of Murtaugh. Dressed as pirates with eye patches and rubber knives, they rang doorbells and returned with bags full of candy. When I dropped everyone off, Matt asked to stay overnight at JR's house. As long as Cathy didn't mind and the children had no school the next day, I was okay with the arrangement. As I drove into the garage, I prepared myself for another hassle. Mike waited for me in the kitchen, arms crossed, his legs planted far apart. I could tell our conversation was going to be intense.

"After you left," Mike said, scowling with obvious annoyance, "I fell asleep on the couch. Before I knew it, that damn pig was in my face!"

"What?" I exclaimed. "She got out of her pen?"

"Yes! She slobbered wet snot on my face. Then she snorted in my ear!"

"I can't believe it," I said in astonishment. "How'd you get her?" I asked, remembering how hard it was for Matt and me to capture her the week before.

"Once I was fully awake, I realized what had happened. I yelled at it and chased it around the living room and into the dining room," he said. "It kept squealing and then it ran back to the laundry room."

Imagining this large man being awakened by a slobbering pig and having him chase her around the house was too much. I dropped to my knees, collapsing on the floor, howling with laughter. Mike didn't think it was one bit funny. But he couldn't help it. His lips started to quiver and then a smile came over his face.

"You have to admit, Mike, that was really funny," I said.

Polly left the laundry room the following day. Faint rays of light fell into the barn as I introduced the horse and pig to each other. The little piglet immediately bonded with Smoky, imprinting as it's called. She soon considered herself a horse, shadowing the large black animal wherever he went, grazing right beside him. Smoky, however, did not like this tiny intruder one bit. When I initially carried Polly into the barn, Smoky stretched his neck to stare at the little black ball. One sniff and he snorted, flattened his ears, and shook his head with disapproval. Eventually Smoky adapted to Polly but he never really approved of her.

The morning sun was already high in the sky a few months later when I carried a pail of leftover food scraps across our driveway and out to the barn. Polly and Smoky were in the pasture, grazing together. Whenever Smoky moved to a new section of grass, Polly embraced his every step. Smoky stood a good four feet higher than Polly, but as far as the pig was concerned, they were twins, adjoined in a forever embrace. Smoky let out a soft whinny as he saw me approach.

"Come on, you two. See what I have," I yelled to the black pair.

Polly snorted and grunted and ran as only pigs do, her squat little legs rushing back and forth as she raced on her toes toward me. As soon as she reached my side, she pushed her nose forward and gave a huge sniff, leaving sloppy snout marks across my irrigating boots. Because a pig's eyesight is so poor, it identifies objects by smelling. In France, pigs are used to sniff out truffles three feet below the ground. Not being a gourmet chef, I discovered truffles are a tuber, a delicacy used in cooking. And one I had never used, or even seen.

"Okay, Polly. Here it is," I said as I upended the bucket, the day's food scraps cascading into her trough. She squealed with delight and dug into a pile of bread, lettuce, potatoes, and meat. Polly was several months old and growing fast. She was bigger than Jack and weighed close to a hundred pounds. Trying not to show favoritism, I gave Smoky a handful of oats and a flake of hay.

After another year had passed, Polly became our second guard animal. Besides being greeted by a black Lab, any house guest would have a large black pig placing her nose at the same height as their side window. Because we had an electric fence that we turned off once Smoky knew his boundaries, Polly would slip underneath and approach all autos coming into our driveway. No way would anyone want to tangle with her as she snorted at their car window.

Whenever I checked on Polly, I always wore irrigating boots. Any exposed area became covered with slimy pig snot. In the evening I brought food leftovers while I cleaned out the stalls and put in fresh straw. Using a pitchfork, I stabbed the straw bale, ducked under the wire fence, and heaved straw into the barn.

Once Polly finished rooting in her trough, she waddled into her stall and lay down. I'd talk to her, using baby words, and told her stories. She seemed to listen as she cocked her head and looked at me. When I sat beside her, cross legged in the straw, I massaged her body. She loved it. Polly closed her eyes, tiny in the massive block of her head, and made pleasant moaning sounds and rolled on her side. I rubbed her ears and noticed how soft they were, as "soft as a sow's ear," an expression I'd heard for years and definitely realized was true. Smoky stood in his stall, eating his hay and oats and glancing periodically over the railing between the two stalls. I told Polly about my day's activities and continued to stroke her body. Pigs don't have fur like most animals, but have dense individual hairs, coarse and bristly, protruding from their rough hides. They also don't have sweat glands. To lower her temperature, Polly would lie in muddy holes to cool the surface of her thick skin. As Polly grew, I asked Dr. Monroe to examine her.

"Should I have her spayed?" I asked, thinking she was like a dog or cat that needed to be neutered.

"Why? Are you expecting some boar to come into your yard?" he asked, almost laughing as he answered.

"No, I guess not," I said, feeling foolish and knowing I had made another "city" comment to our country vet. Thank goodness, Bob had a lot of patience and enjoyed his visits to see us.

* * *

Later the following summer I asked Matt if he wanted to ride to the gravel pit. I inserted a metal bit into Smoky's mouth and raised the bridle's headpiece above his ears and pulled his forelock above the browband. As usual, I wore a riding helmet, boots, and half chaps.

"Sure. I'll get my bike," he said with a smile.

I swung the English saddle and pad onto Smoky's back and reached under his belly to tighten the girth belt, buckling its two saddle straps. Once I undid the fence gate and mounted Smoky, Matt jumped on his blue bicycle and pedaled away. He called to Jack as he raced down the driveway. Polly, wearing a bright pink collar, followed the three of us to the front yard. As we started toward the dirt road, I heard a meow and turned to see Buddy, our black cat, rushing to join us.

By that point, we had owned Polly for three years. I knew she no longer thought of herself as a pig. She had imprinted on Smoky. She weighed over four hundred pounds and trailed Smoky everywhere, chomping on tufts of grass right beside him. Smoky and I guided the impromptu parade with Matt and Jack beside us. Twenty feet behind were Polly and Buddy. We slowly progressed the mile distance from our driveway to the gravel pit.

I eased into the rhythm of Smoky's walk as we passed huge boulders stacked on the side of the road, leftovers from when the original ranch had been carved from the prairie. They were the size of small cars and this was the exact place where Smoky often spooked. It seemed to be his annual ritual to attempt to buck me off and another reason why I always wore a helmet. As we rode by the boulders, I squeezed my knees together and turned Smoky's head to look directly at the rocks. He relaxed and our parade resumed. The dirt road was pitted with rough stones, and it became increasingly difficult for Matt to balance his bicycle. The bike wheels shifted every time he pedaled on the irregular ground.

Just as we reached the junction to the gravel pit and were beginning to turn left and venture up the trail, Matt's bike skidded on the uneven road. He fell to the side and his body crashed to the ground. Jack barked several times and Smoky jumped straight into the air. When Smoky's hind hooves slammed into the dirt, one of his feet missed Matt's head by less than an inch. It was so close I saw no separation between his bone-crushing hoof and Matt's head. Matt got up and shook himself off, not realizing how close he had come to being maimed or killed. I dismounted and walked to Matt, checking his head and hair. Everything looked good. He wasn't hurt, just a few scratches. With tears welling in my eyes, the incident reminded me once again of the constant danger of living around large animals. Even if there had been no intent.

"Let's go back," I said as I placed my foot high in the stirrup and pulled myself onto the saddle. "Your bike can't handle this rough trail. And Polly and Buddy can't keep up."

"Okay. I'll race you back," Matt said as he jumped on his bike. He was off, churning up a haze of dust with no thought of his possible injury. Jack ran right beside him. Smoky and I followed along with the last two of the parade trailing far behind. By the time we reached our driveway, the group had scattered along the length of the road. I shortened my reins and stopped Smoky, twisting in the saddle to make sure the stragglers made it back to our property. Once inside the electric fence, I dismounted and closed the gate. After removing his saddle and bridle and returning them to the tack room, I rewarded Smoky with a carrot.

"Wash up, Matt. Dad will be home shortly," I said and walked into the kitchen to start our lunch.

Chapter Thirty

Hailstorm

"Hey, Mike. I'm going to photograph the combines today. The sky is perfect for photos. Clear."

"Okay. Bring Matt. I'm in the field behind the Adams' house. Clear."

With harvest in full swing, the late August weather turned hot. The day before had been a scorcher, a blanket of searing heat. I stood on the back of Mike's pickup and watched, nearly blinded by the sun, as the men crossed the field in Harvester combines. My Nikon camera was out and ready. Four combines moved through the grain, one behind the other but in separate rows. A dark navy sky shadowed the background while the red combines crossed the yellow field in the foreground. Streaks as light as butter, smooth and white as candle wax, graced the grain stalks. This was the photo I had been waiting for.

The colors were crisp, clear, and bright. I took numerous shots, and then jumped from the bed of the pickup and climbed into the cab. I drove from the field and back to our home as grain shafts danced in the wind, and the men in the combines continued to harvest. Soon clouds gathered, crowding the sky and shutting out the sun. Back at the house I heard a low rumble of thunder that heralded the approaching storm.

And then it came. Hail the size of marbles burst from the sky, pounding the ground and beating against our house like gun shots

at a rifle range. For fifteen shocking minutes, devastating ice balls inundated Sky Ranch. I stood in the dining room as the blackened sky moved across the horizon, blowing swiftly to the east. Then it stopped as if a spigot had been turned off. Sun fought to break through the angry clouds, and Mike, Matt, and I drove out to witness the damage. Whole sections were flattened and destroyed. Weather always played a huge part of their income. Sky Ranch's crop and its ensuing money had, today, been damaged in a flash.

Mike and his brother, Don, could do everything right, plant and harvest at the right time, enjoy high future stock market prices, and then have crushing weather conditions. I never knew how tenuous their occupation was. Once I moved to the ranch, I became exceptionally aware of the weather and its ensuing results.

The next day I passed Mike on 4900 going in the opposite direction. We stopped and he rolled down his pickup window and rested his shoulder against the door.

"Hey. Where're you going?" he asked.

"I'm on my way to Twin," I answered. "Going to buy groceries. Do you need anything?"

"Not now. Call back before you leave."

"Okay," I said as I started the engine and pulled away from his truck, driving north on 4900. Looking in the rearview mirror, I saw another pickup pull into the spot I had just left. The two men yakked across the span of their trucks, maybe chatting about specific farm issues but most likely talking about the hailstorm. Who knew how long they sat there, parked in the middle of the road? They certainly didn't have to worry about traffic. In a full day, maybe a dozen vehicles drove over the road. We really were in the middle of nowhere.

Chapter Thirty-One

The Little Farm in the Canyon

"Want to buy some property in the canyon?" asked one of Mike's fishing friends from Sun Valley.

"I don't know. Tell me about it," Mike requested.

"We used it as a hunting and fishing retreat. It's located on the Snake River down in a canyon between Filer and Buhl. The hundred and fifty acres was once a cherry orchard. I think in the 1930s."

Mike and I checked out the parcel the very next day. Most of the fruit trees were dead and looked like dark, gaunt ghosts, misshapen and protruding from overgrown weeds. It was covered with hundreds of Russian olive trees, and the land had been left fallow for at least thirty years. Because the parcel bordered a half mile of river, we saw the potential as a small residential development and decided to buy it. Mike and I still lived at Sky Ranch but now he and I owned some land by ourselves. We called it the "Little Farm."

Once the purchase had been completed, I realized what farming really entailed. Instead of a large home at the ranch, we lived in a twenty-foot trailer with no running water and no bathroom. Granted, we were only there for long weekends but that was enough.

The first year of ownership was an exciting time of discovery but it practically did me in. I felt like I was living in a foreign country. Working a farm was so unlike anything I had ever experienced. Certainly nothing like my youth at country clubs and private beaches. Not only was I making three meals a day but I was also hand-irrigating recently planted trees and removing trash, weeds, stones, and stumps from our newly purchased land. Matt and I, along with a few of his buddies, trudged from the Snake River carrying five-gallon buckets to water twenty Lombardy poplars planted as a windbreak. We smelled the dampness as we moved through the wetlands bordering the river, dark mud sucking at our boots. As each day faded, I bent at my waist, my hands on my knees, and just breathed. Irrigating by buckets sapped my stamina. Sweat ran down my face and stung my eyes. It was hard work. Local kids, even though they were paid, usually lasted only a day or two. Matt and I didn't have a choice.

Mike and Mario alternated digging ditches with Sky Ranch's backhoe. They formed numerous trenches, diverting part of a stream flowing from a canyon waterfall and ran it through the upper part of the Little Farm. Because we needed to irrigate our land, we purchased water rights when we bought the property. Just because a stream ran through our property did not mean we had the right to divert it. We had to purchase the "right." Once the water started flowing, the place began to green, and struggling fruit trees once again flourished.

Whenever there was a moment to spare, Mike, Matt, and I removed rocks, rubbish, and loads of dead trees. We put the wood into a pile for the evening's hotdog roast. The rest was thrown into a giant cavity Mike had dug with a backhoe. We found furniture, tires, cans, and seemingly miles of barbed wire. All were flung into the deep pit and burned. Using a tractor, Mike covered the hole with four feet of dirt and leveled the spot. He planted corn over it the next year. But not everything was discarded into the deep cavity. We saved a 1930s car frame and an oak-wheeled wagon. Mike plowed a new entrance road around the two antiques, and I planted white and yellow daisies close beside them. The Little Farm began to take shape.

Then we tackled the endless Russian olive trees. Although beautiful, grey-green trees, they multiplied like weeds and had sharp, three-inch-long thorns protruding from every branch. No longer a debutante wannabe, my job was to crawl underneath the lowest branches, hanging about a foot from the ground, wrap a metal chain around the trunk, and crawl back out. My jeans and chest chafed against the ground as I wriggled back and forth; my battered body taking the brunt of the chore. Matt grasped the end of the chain from me and handed it to his father who then attached it to the frame of the backhoe. Matt and I stood far to the side while Mike climbed onto his machine and put it in reverse. Once he pulled out the tree, he dragged it to a growing pile of shrubs, piled haphazardly like Lincoln logs that were burned the following day. We worked many hours and even into the evening. The backhoe had strong headlights, and we could see long after the sun had set.

During one particular night, I slithered under a Russian olive tree, gripping the chain in my right hand and fought through a tangle of thorn-covered branches. The needles on the ground ripped at my jeans and tore my T-shirt. Hemmed in, I reached out to wrap the metal cable around a trunk. As I stretched toward the tree base, a branch snapped at my face. The nasty thorns dove into my cheekbone and punctured my skin just below my eye. I stopped and backed out on my hands and knees. Mike saw me in the tractor's headlights and realized I didn't have the chain. He jumped off the machine and came over to me. Noticing drops of blood sliding down my face, he wrapped an arm around my shoulder, and I wept into his chest. I was exhausted.

"Okay. That's enough," he said. "Let's go home."

He turned off the machine and the three of us walked, filthy and drained, back to the trailer by the glow of a flashlight. I could barely stand but I opened two cans of chili, added chopped onions and grated cheddar cheese, and heated our meal. Saltine crackers and glasses of milk supplemented the dinner. Dirty as we were, we fell into our beds and were asleep within minutes.

By the next year, Mike had a waterline and septic tank installed. The Angler's Company purchased a one-bedroom RV with a living room pop out, a bathroom, and a mini kitchen. This was heaven compared to the year before. We pushed hay bales against the outside foundation, insulating the trailer from both the searing heat and the freezing cold of the canyon floor.

Friends and family dropped by to help with chores and to partake in our evening campfires. They'd gather fallen branches, knock over dead trees, and collect wayward rocks. After several hours of work, we'd sit around the bonfire with sodas, a beer or two, and talk about the day's activities. I'd bring out a plate of hot dogs, buns, and potato chips. We took sticks and speared the dogs lengthwise, cooking them until they were slightly burned. At the end of the meal, we often had s'mores. Mike's grown children, Christa and Blaine, as well as our Twin Falls friends, Carol Roseberry and Del Carraway, were our normal guests. But we never really knew who would drop by. Sometimes fishing and hunting friends and sometimes our Friday-night bridge group. We always seemed to have a crowd during our weekend roasts.

*   *   *

In spring at the Little Farm, we'd wake to leaves casting a light green tint with poplars and cottonwoods budding as they warmed. Shards of sunshine sliced through the tree tops and warmed the canyon floor. While Mike and Matt handled early-morning chores, I stayed inside the RV and completed breakfast. By the time they returned, coffee brewed on the stove and the aroma of bacon filled the air. They settled on the couch and pushed bacon, eggs, and toast into their mouths at record speed. Before long, their empty plates indicated all was good.

"Look what I found," Mike said as he held up a black pointed stone. "It's an Indian arrowhead."

"Can I have it?" Matt asked as he reached for the sharp object.

"Sure," Mike said. "But be on the lookout. This canyon was a perfect place for the Indians to live. I'm sure you'll find many more." And so we did. Whenever the soil was turned over, getting it ready for planting, Matt and I would scour the land, looking between the dirt rows to see if we could find some more Native American keepsakes.

"What'd you think about getting a mule?" Mike changed the subject and asked me. A toothpick wagged from his lips as he spoke.

"Why would I want a mule? I have Smoky."

"I'm talking about a machine. It's called a Mule and it's made by Kawasaki," he stated.

"Can I work it? I don't even know how to operate your motorbike," I said.

"We'll drive by the dealership on our way home," Mike suggested. "We'll see."

It turned out to be the best present I had ever received. I loved the mechanical Mule. It had four wheels, headlights, a bench seat, and an open cargo bed with a tailgate. Instead of carrying buckets of water, I could place a huge plastic tank in the back, fill it with water, and maneuver around to the windbreaks and the surviving apple, cherry, and pear trees.

During early summer afternoons, Matt and I took the Mule along the dirt road bordering our property and gathered low-hanging apricots and peaches, branching out from Kelley's Orchard, its fruit dangling over the roadside fences. We then scrounged our own pastures and picked wild asparagus. If we harvested several bunches each weekend, the asparagus kept growing. We usually had a good crop through August and we never had to plant or weed them. When we returned to the RV, Matt stuffed a fistful of spears into a water glass. Once again, we had fresh asparagus for dinner.

Alongside the flat field on the western edge of our property, lay a shallow pond loaded with marsh grasses. We heard clacking calls from migrating birds: redwing blackbirds, bobolinks, and yellow-headed blackbirds graced the swampy land. They perched on cattail tops, their festive feathers shining in the sun, looking like a just-opened box of crayons. A few times I even heard the melodious sound of a meadowlark, its flutelike voice would stop me in my tracks as I turned and listened to its song.

All across the canyon land, crow-sized magpies swooped in and out of trees and rested on gigantic boulders. In Idaho, the magpie is stunning in its tuxedo colors: its head, chest, back, and tail in vivid black, intensely contrasting against its stark-white body and dark turquoise wings. The bird was not welcomed by duck hunters because it consumed so many waterfowl eggs. They would shoot the beautiful magpies whenever possible. I discovered besides eating eggs of other birds, the magpie is also an opportunist. It feeds on ticks, rodents, carrion, and garbage. I enjoyed the attractive bird as it consumed the repulsive items I didn't like. I, once again, chose the naturalist's approach.

While at our Little Farm, I never went near the Snake River without smothering myself with bug repellent. The marshy area was alive with mosquitoes. The dense trees shut out light and welcomed these blood-sucking insects. I waved at clouds of bugs but they always seemed to find me. It was as if their scouts would locate me and report back to mosquito headquarters. I was fresh meat. They whined near my nose and ears and my vision warped into a veil of black specks as they engulfed me. In the heat of the day, they drank my blood and sipped the sweat from my back as I bent to pick fiddleheads for our supper salad. By the time I returned to the trailer, I was covered with welts.

Besides mosquitoes, the thick grasses embraced an abundance of ticks. In the trailer, we checked our skin and hair every night. One time we found over thirty ticks imbedded in Jack's body. What a difference from the bone-dry land at Sky Ranch. The Little Farm, situated so close to the Snake River, produced many more insects and animals. It was home to skunks, opossums, chipmunks, squirrels, rabbits, rodents, snakes, mule deer, and rock chucks (or yellow-bellied marmots). Although rock chucks are roly-poly cute with thick fur and big teeth, they are prolific reproducers with few enemies. In one year, a male could impregnate several females, and each mother could produce four pups. That's a lot of babies! Because their diet consisted of grasses, they loved our alfalfa fields. During summer the second year, we lost three acres of alfalfa to these bountiful rock chucks.

One afternoon, Matt and a friend from Buhl High School arrived at the Little Farm loaded with guns and ammunition. They crept to the top of a ledge and shot at the animals as they ate through our alfalfa field. Once the shooting started, they heard the rock chucks whistling their distress, warning others of eminent danger. The two boys did a good job. We had fewer rock chucks every time they came to shoot.

When Bill Lemmons from B & B Apiaries heard of our newly planted alfalfa field, he asked about putting some bee hives on the Little Farm. "I'll even give you a couple half-gallon honey jars."

And how delicious the honey was! I soon cooked with honey instead of sugar and our desserts and breads became sweeter in the process. Between the fruit trees, ducks, geese, trout, and products from Sky Ranch, we were truly living off the land.

As River Road became more paved, a nearby rancher added a metal cattle guard at one end of his property and a fake one at the other end. The yellow stripes painted on the road mimicked a real cattle guard. Surprisingly, the cows knew not to cross it. With their slobbery noses, hairy ears, and curious nature, they'd approach and smell the yellow lines, but would not dare crossing the painted imitation.

During the following summer, I drove from Sky Ranch, past Filer, and turned north toward the Snake River. My car spewed a cloud of dust as I meandered down the canyon road and approached the wire entrance to the Little Farm. Mike and Matt had trailered Smoky to the canyon an hour before me and were already at the RV, waiting for me to arrive. Having my horse in the canyon all summer long was exciting. I couldn't wait to explore the hidden niches of the Little Farm from Smoky's back.

A few days later, Scott and Brad Traxler, young sons of our canyon neighbors, joined Matt and me standing next to our RV. "Want to play hide and seek?" I asked. They had the whole canyon as a playground, and yet they were stymied as to what they should do next.

"Sure," they said in unison.

"We'll stay on this side of the road," I instructed. "There's plenty of places for me to hide. You'll never find me!"

"Oh, yes, we will," Matt said. "I know all your hiding spots."

I saddled Smoky and told them to keep Jack. Sixteen-year-old Matt rode my Kawasaki Mule and ordered Jack to sit on the floor boards. Brad and Scott each had their own four-wheelers, all-terrain vehicles that are similar to motorcycles but with four wheels.

"Count to one hundred. Then try to find me," I shouted as I aimed Smoky toward the western end of the property and loped down a dirt trail. Once at the far end, I turned toward the river and wove in and out of tufts of high grass and around green ash trees and willows. The thick leaves billowed and flapped like sails in the wind and for the next hour, they looked for me. With all of them yelling back and forth, I knew exactly where they were at all times. Sitting on top of Smoky, I also had a height advantage. I avoided the boys and moved silently from one area to another. I came out of the marshy woods and cantered through a nearby field, the sun shining on my face. A large mule deer emerged from the shadows and darted, fleeing deep into the trees beside the river. I waited for another to follow and then we cantered back to the RV. Talk about a playground. You couldn't ask for a more spectacular place, deep in our canyon with waterfalls cascading on several sides and fruit trees intermingled with huge boulders and flowing ditches. The boys never did find me, but they were hungry and soon returned. It was, once again, time for hot dogs and s'mores.

On our third year at the Little Farm, Mike leased part of our land to a local farmer who trucked in about fifty head of cattle. Every few weeks, Matt, Mike, and I moved the herd from one field to another, a procedure known as "rotational grazing" as it kept the grass fresh and nutritional. Mike took his four-wheeler, Matt had my Mule, and I rode Smoky. He reverted to his expertise as a cattle herder and knew exactly what to do. He positioned himself to ward off any stragglers and drove stray cows back to the main group. Because he turned and twisted so quickly, I kept my seat by squeezing my knees around his trunk and tilting my body with his every movement. We were, once again, a team, and I was in heaven.

Since the Little Farm now had cows on the property, our weekends became a battle between us and massive hordes of menacing flies. Clouds of big-eyed black flies beat at our RV's screens, eager to get inside. We had to close the trailer's door as fast as possible; otherwise, a stampede of black flies charged inside and inundated the interior.

As the summer waned and autumn approached, the flies and mosquitoes diminished. We prepared our land for winter, doing necessary farm chores, all new to me. Mike taught me to drive an old 1950s Farmall tractor that was small and simple to operate; I could easily shift the gears and turn the front wheels. We worked together, burning ditches to get rid of the weeds growing on the sides of the dirt road bordering our property. While I steered, Mike held a long metal rod and torched the grass and weeds along our fence line. He walked behind the tractor and to my right while I maneuvered the vehicle to make sure I did not extend the rod beyond his grasp. I was a true farmer's wife, working the land right beside my husband.

At the end of the season, it came time to load the animals for market, and a cattle truck backed into the far corner of our property. We grouped the herd into a temporary enclosure near the rear of the large, grey vehicle. Once Mike opened the gate, the cows hesitantly climbed the truck's ramp, banging the metal incline with their heavy feet, bawling and protesting with angry kicks as they fought for passage through the loading chute. Mike whistled, the trucker and Matt waved their hands above their heads, and I sat on Smoky watching the process. Eventually, the cattle were on board and the trucker slammed the back doors shut. The heavy vehicle slowly drove away from the pasture and made its way down River Road toward Buhl.

The next day Mike borrowed a neighbor's phone and called back to Sky Ranch. He was told that one cow had not joined the others on their journey to market. Checking off identifying ear tags, the owner knew he had not received one of his cows. Matt joined Mike in his pickup while I rode Smoky. We searched the field in which the cattle had last been pastured and located the cow. She had hidden herself in a thick circle of trees, so dense you could only creep beneath the lowest limbs to reach the center.

"You're small," Mike said to Matt as they stood beside the pickup, assessing the situation. "Why don't you crawl in and drive her out?"

"Okay," Matt said as he walked toward the trees. He looked back at us and then bent to his hands and knees and pushed through the bottom branches. At no time did Mike or I think this was dangerous. We just thought Matt would scare the cow and she'd leave the comfort of the trees. As it turned out, she let out a huge bellow and stomped her feet. She was not going to leave her sanctuary. Matt edged around behind her, yelling and throwing stones in her direction. Eventually, she staggered from the impenetrable thicket and appeared in front of Mike and me. There stood an angry Angus cow with a newborn calf.

We never knew she was pregnant. No wonder she hid from the other cows. Mike opened the tailgate of his pickup and pulled down a wooden ramp. We circled the mother cow and encouraged her to climb the ramp. By this time, Mario had joined us and she succumbed to the four of us prodding her into the pickup. Her calf followed. Angus cows are fierce mothers. Matt had been pretty vulnerable, trying to move her from the thicket. We were extremely lucky she hadn't attacked him.

When autumn tumbled into winter at the Little Farm, and withering leaves fell to the ground, duck hunting began in earnest. We rebuilt an old shooter's blind that had been made of scrap wood, its roof covered with stalks of brown reeds. A small bench remained inside. The blind opened on three directions, the largest of which faced the Snake River. Surrounding the blind; cottonwoods, willows, and ash trees grew close together, their drooping foliage touching the water between dense clusters of bushes. This was the perfect blind for our weekend forays at the Little Farm.

In our RV Mike and Matt arranged their gear ahead of the morning hunt. Before the sun reached the canyon, they were awake. I gave them cinnamon rolls and made a thermos for each, one of hot chocolate, the other of coffee. They gathered their equipment and covered themselves with heavy jackets and wool caps. Jack had already bounded down the trailer steps, jumping with anticipation. Still in my nightshirt, I peeked through the curtains as they walked away. Soon they became dim shadows in the early morning light and I returned to bed and crept under the covers.

When the sun rose higher in the sky, I ambled out of bed and changed to a turtleneck, sweater, jeans, and sneakers. I pulled my hair back into a ponytail and added a baseball cap. After a cup of coffee, some toast and orange juice, I left to find my boys. I listened to the rush of the Snake and finally heard shots in the distance and headed in that direction. They had left the blind and had taken a boat with all their gear to a little island about twenty feet from shore. I couldn't see them but I heard duck calls echoing across the canyon. From there, I turned to clean a nearby field of fallen sticks and checked on some newly transplanted trees. A bird squawked and departed as I approached. An hour later, we joined at the RV, and I prepared a hearty breakfast of cereal, orange juice, toast, bacon, and eggs.

"How'd you do?" I asked while I set the table and handed Mike a mug of steaming coffee. I placed juice glasses near the plates and gave the toast a light buttering.

"I shot one mallard, that's all. But you wouldn't believe what happened," Mike said. "Once we were on the island, I showed Matt how to use a duck call. But when I later checked, our boat had floated away."

"Did you get it?"

"Yup, but what a hassle. I had to cross to shore, walk down the river's edge, and finally catch it," Mike said. "But that wasn't the worst part. I had to drag it against the current, back to the island. It took forever!"

"How'd you do, Matt?" I asked.

"Great. It was fun with Jack. If I had shot one, he would have been right on it."

"Yeah," Mike said. "I was worried about them. I was afraid Matt would try to cross to shore since I'd been gone so long. The current there is really swift."

"Good for you, Matt," I said. "Good that you stayed on the island."

"Yeah. And as I retrieved the boat," Mike said with a laugh, "I heard the most God-awful sound coming from the island. No way a duck would respond, but at least I knew Matt was safe."

They continued hunting for the rest of the season and although Matt never accomplished the intricacies of duck calling, he did get in some good shots. Between the two of them, we always seemed to have plenty of ducks in our freezer back at Sky Ranch.

Chapter Thirty-Two

Fly Fishing Henry's Lake

After spring planting and before harvest, work at Sky Ranch slowed to a crawl. Employees scheduled vacation days, and Mike booked a week at Staley Springs. The rustic facility had a dozen cabins and sat on the west side of Henry's Lake, fifteen miles from Yellowstone National Park.

We trailered our fourteen-foot Mirrocraft boat to Interstate 86 and drove past Pocatello. At Idaho Falls we veered off and continued north on Highway 20, turning left toward Henry's Lake. Even in summer, the highest peaks of the mountains bordering the shallow alpine lake were capped with snow. Five hours after we left the ranch, we arrived at the vintage resort. As soon as we parked beside our designated cabin, Matt and Mike claimed their fishing gear from the back of the truck and ambled toward the lake. They walked away from the cabin, each carrying a float tube, their life vests, and a fly rod. Matt used my old Fenwick rod while Mike brought his brand-new Sage rod. This was my time to relax. Jack followed me around the log cabin while I placed clothes in dresser drawers and cans of food in knotty-pine cabinets. I stocked meat, vegetables, beer, and sodas in the refrigerator and then tackled the fireplace. After placing kindling in the metal grate, I brought in logs from outside. I was nesting. It was what I liked to do.

The boys wouldn't be back for a couple of hours. I braided my hair and dressed in a turtleneck and jeans. Once I laced my hiking boots, Jack and I ambled up the mountain slope behind the lodge. We passed wide meadows teeming with butterflies and colorful wildflowers. Birds squawked and swooped around us, alighting on nearby branches, staring at us. We were the intruders in their space. I turned my face upward and soaked in the late summer sun. It felt good to be free in the woods, to feel happy, to smell the pine-scented air. We climbed higher, walking over a carpet of pine needles, as squirrels and chipmunks gathered nuts for winter. In the distance, I saw cattle grazing in fields beyond the lake and heard a myriad of birds, calling from trees high above our heads. Jack wagged his tail and continued to move up the hill, his nose to the ground.

The sun shone through the aspens, creating butter-like colors whenever the leaves shimmered in the wind. I heard rustling from the trees and marveled at the white tree bark below their canopies, the aspens reminiscent of dogwood trees back in Connecticut. I was at peace. Soft cool breezes heralded the arrival of autumn just a month or so away. I turned and retraced my steps, heading down the mountain. As we hiked through a meadow, I picked a handful of blue and orange wildflowers, reminding me of my mother who always had fresh flowers in vases throughout my childhood home.

When I entered the cabin, I opened curtains and brought in as much light as possible. The main room was a combination kitchen, living, and dining area. At night Matt slept on the couch with Jack sprawled at his feet. Mike and I had a small bedroom next to the bathroom. Near the entrance, a Formica table rested under a picture window facing the lake. I started dinner and arranged the wildflowers in a glass jar. By the time Mike and Matt returned, the low slant of the afternoon light had arrived. I had a meal prepared and a fire blazing, crackling in the grate.

Once they had washed and sat for dinner, I heard about fishing under the twenty-foot-high "wire." Beside the main resort building, a stream dumped into the lake and flowed past the wire. It was illegal to fish between the overhead metal line and shore, thereby giving trout a chance at survival. In their float tubes, they maneuvered as close to the wire as possible but only caught and released one cutthroat during the afternoon.

"Del and Carol were there along with a half dozen other anglers. We all had good luck. But tomorrow, we'll take the boat," Mike stated.

"Where're we going?" asked Matt.

"Down the lake, toward the campground. We'll fish about a mile from here."

"Will you be home for lunch?" I asked.

"No. Just pack something and we'll stay out all day."

Yes! I thought happily. *Another day to myself.* I would have a chance to identify birds and wildflowers and to hike around Staley Springs with Jack at my heels. It was my time to relax, to remove the pressure of ranch life and the Angler's business.

There was no television, or even a telephone, in the cabin. When we gathered at night after dinner, it was just the three of us enjoying the evening. As flames spit up the chimney, Mike placed his Renzetti vice at one end of the dining table and tied several dry fishing flies. He took a Coors can from the refrigerator and flipped open the tab. After he took a deep swig, a frothy white cap covered his mustache. He wiped it off with the back of his hand and stared into the flames, watching a blaze move among the logs. I brought out a deck of cards and Matt and I began playing "kings in the corner." We filled the room with jeers and laughter as the game progressed. Shadows danced on the walls and Jack fell asleep on the floor. He turned in his sleep and wriggled himself onto his back, his feet in the air. His front paws flexed at the wrist and twitched as he dreamed.

When the fire began to die, I stirred the logs with a poker and added more wood. Soon the clash of red embers lit the remaining wood. We played cards until the wee hours while Mike tied fishing flies. That night was typical of our evenings at Henry's Lake and it continued to be so for the rest of the week. Relaxing and fun.

The last day at Staley Springs meant a morning of fishing for Mike and Matt while I packed. I kissed them goodbye before they walked to the boat dock, loaded with fishing tackle, a thermos, and a camera case. I turned back to complete the cabin chores and just as I began to empty the refrigerator, the front door burst open. Matt charged inside, swaddled in a life vest, and dripping wet.

"What happened?" I yelled.

"The boat flipped," he said while standing in front of me shivering. "I'm freezing!"

I grabbed him by the shoulders, steered him toward the bathroom, and turned the shower knob to full force. Matt pulled the curtain open and stepped in under the powerful stream, clothes and all. He lifted his head and let the steaming water fall over his body.

"Where's your Dad?" I asked through the plastic curtain.

"He's getting our boat."

Just then, Mike walked in, soaking wet as well. I ran to the bedroom and grabbed a blanket and threw it over his shoulders. He wrapped it around his body, shaking and stomping his feet. I rubbed his back and arms, trying to warm him.

"What happened?"

"Matt asked to steer the boat," he answered as he continued to shiver. "He's been around boats for years, so I said yes."

"So, what happened?"

"He was doing fine. Then he started to go to yesterday's fishing spot. I pointed to another area across the lake by the old lodge. But he shifted the tiller too fast. In a second, the boat flipped and we were in the water."

"How'd you get here?" I asked as I continued to rub him.

"Our boat made a circle and was coming right back at us. I thought it was going to run over Matt. It was heading straight toward him."

"What'd you do?"

"As it flew past, I grabbed its side and pulled myself up. I don't know how I did it but I did. I hit the 'kill' switch and the engine stopped. Just then, two boats came over and got us. They picked up everything that floated. The rest is at the bottom of the lake."

"I'm glad you're safe. And our boat?"

"One of the guys hooked it to his. He towed it in and I dragged it on shore. We can handle it later."

"Matt's out of the shower. Why don't you get in?" I suggested.

Matt and I finished packing. I cautioned him, saying, "I'm glad you were wearing your life vest. You never know when disaster can strike."

Once Mike loaded our boat, we were on the road, heading down the mountain, past the Tetons toward Sky Ranch.

Chapter Thirty-Three

Twin Falls County Fair

I t was the Friday before Labor Day, known as East End Day at the Twin Falls County Fair. All schools, banks, and businesses in East Magic Valley shut down for that one day.

As we walked from Mike's pickup toward the fair entrance during East End Day, a steady stream of spectators surged past us and descended onto the fairgrounds. The local weatherman predicted rain but it made no difference. We never missed a day at the fair. A crowd flowed from vehicles that had been parked, row after row, in a nearby field. It seemed everyone was involved—either as participants, families, sponsors, or volunteers. Murtaugh schools had closed for the week since so many of the town's children were involved in 4-H.

I wore dress jeans and a pair of coffee-colored, lace-up leather boots. Over a white shirt, I had on a navy fringed suede vest. I wrapped a red scarf around my neck and pulled my hair back into a ponytail. That was as close to looking like a cowgirl as I could manage. In the meantime, Mike had polished his dark brown snakeskin boots and wore dark jeans with a brass competition belt buckle he had won at a team-roping contest. The white Stetson completed his look. Matt wore jeans, a plaid shirt, and Nikes, typical of his farming friends. The three of us worked our way down the midway, past the games and concession stands, to the animal buildings at the rear of the grounds. In the sheep barn we smelled dirt, animal droppings, and wood shavings. But the stalls were empty. Everyone had gone to the judging ring. For months, our nieces, Gina Dawn and Sarah, had walked their sheep along Sky Ranch roads, teaching their baby lambs to stand and heel until they became forty-pound sheep. Surprisingly, both girls were in the stands, not in the arena with their sheep.

"Where's Candy and Woolly?" I asked as I stood beside Georgina. Gina Dawn and Sarah stared at their feet and never said a word.

"The girls washed and combed the sheep this morning and then they tied them to the front fence," she said looking at her daughters. "They needed to take showers and change their clothes before we left for the fair. When they came back outside, both sheep were dead. They strangled themselves on their tie ropes."

"I can't believe it!" I exclaimed. "My gosh, they've worked for months with their lambs!"

"It's really sad," Georgina said, looking as if she were going to cry. "Sheep are known for being stupid, but this was incredible."

"I'm so sorry," I said to the girls, leaning down and touching their shoulders. After pausing a few moments to take in the tragedy, I asked, "Will you be barrel racing later?"

"Yes, we'll be there," Gina Dawn and Sarah said in unison as they lifted their heads and tried to smile.

Mike and I sat next to them in the bleachers as a series of children guided their sheep into the dirt arena. They were as clean and polished as their animals. They held tight to the halters and moved their animals in a large circle, either by tugging or coaching. It was a tense time for both the youngsters and their parents. They had worked all summer for that exact moment. Each child stopped in front of the audience, trying to keep their animal still. Then we heard the bidding begin. Because the money went to the individual child and not a professional organization, the prices reached super high levels. Kay Wolverton, my other sister-in-law, and I bid and won a lamb. We had the meat split between our two families, and I knew it would be one dinner Mike would never consume. Typical of a cattle rancher, he considered sheep unwelcome on grazing lands as they nibbled the grass to its stub, leaving nothing for cattle. He called sheep "range maggots" and he refused to eat them.

Knowing Don had brought their RV to the fair and would be staying the week, we told Georgina we would see them later. We left to inspect the rest of the 4-H sheds and spotted several youngsters washing and brushing their animal charges. Some had cows, others had rabbits, chickens, goats, or pigs. Polly had become quite large, but she was too much of a pet for me to include her at the fair. I showed a picture of her to some of the volunteers in the pig shed. She wore a pink collar when my parents came to visit and joined us on the patio when we sat for lunch. She intimidated my father, but my mother thought she was cute.

From there we moved to the merchant buildings; then on to specific structures, ones that held antiques, art, photography, flowers, vegetables, and baked goods. Judges attached ribbons to the top

contestants; and because Magic Valley was sparsely populated, Mike and I knew many participants and several winners. After checking the displays, we threaded through meandering crowds on our way to the food booths. Volunteers placed salt and pepper shakers, napkins, and the ever-present toothpicks on each table. Winds gathered as we stood in line at one of the Boy Scout stands, and thickening clouds appeared in the distance. Paper plates bent from the weight of our burgers as the three of us walked to a white-painted picnic table and sat down. Winds increased and napkins scattered throughout the grounds. A group of school kids made their way to us and invited Matt to join them at the fun house.

"Can I?" he pleaded and looked at his Dad. "And can I go on some rides, too?"

"Sure," Mike nodded since we knew the boys. The entrance fee to the fair included all amusement rides. We didn't have to worry about forking over any additional money.

"Meet us here at six," Mike instructed. "Listen for the rodeo starting time. They'll announce it over the speakers."

I turned back as the boys raced away. The crowds swallowed them as they disappeared amid a mass of laughter. Mike and I added slices of apple pie to our plates and joined several farmers and their wives at a nearby table. Fair time was that one special occasion when faraway friends, those you don't normally see, come to town and you have a chance to catch up on their families and activities.

As the winds intensified, a veil of dust blew through the grounds. We saw tiny tornadoes whirling between the crowds and then we heard the announcement. The rodeo would start in half an hour. Mike and I made our way down the midway, nodding to passing neighbors and headed toward a knot of people, standing in front of the ticket kiosk. Matt came running to us as Mike stood before the ticket window.

"Can I stay with Kellen and Kirby?" Matt begged. As an only child and living so far from others, I knew it was what he really wanted. "Okay, but ask your father," I said.

"Dad! Can I?"

"Okay. But be sure to be here when the rodeo ends," he instructed.

Matt was gone in a flash, swept away in a crowd of youngsters. The rodeo no longer held his attention as it once did. Now it was his friends who captured him. Mike bought two tickets, and handed me a brochure. Team roping was one of our favorite events as Mike was a former competitor and knew several of the contestants. We sat on a hard bench and shouted support as the competitors completed their routines.

Barrel-racing was another enjoyable event as both Gina Dawn and Sarah participated. They wore cowboy hats and sequined shirts tucked into black jeans. Besides being good looking, they were accomplished riders and crowd darlings. Mike and I watched each one race around three, fifty-gallon drums in a cloverleaf pattern, hugging the barrels and leaning into the curves. As teenagers with shirts sparkling in the stadium lights, they attracted rodeo cowboys, the young men who wore fringed leather chaps and colorful scarves, strutting around the Wolverton RV like parading peacocks.

The setting sun transformed the horizon from subtle pink to an intense gold. Before long, large cumulus clouds jostled against each other and rolled across the darkening sky. Beams of lights continued to brighten the arena and the rodeo continued. Soon the bull riding event began. It's the most dangerous contest in a rodeo. It's an eight-second event, pitting a young man against a bucking, kicking, and spinning bull—one that weighed close to two thousand pounds. The cowboy holds onto a rope tied around the bull's belly with one hand. The other hand is held high above his head and is not allowed to touch the animal. During the 1980s, riders did not wear helmets or padded vests. It was considered cowardly. Today, almost all riders wear protective gear.

Clowns were another important feature at the rodeo. They diverted the charging bull from stomping or even goring a thrown rider. When the bull-riding contest started, I stood and climbed the bleachers to go to the restroom. I couldn't even stand to watch the event. Before long we left the rodeo and gathered Matt. He had waited at the rodeo entrance, his sticky hands wrapped around a paper cone topped with fluffy, pink cotton candy.

"Where'd you get that?" I asked.

"Bill bought it for me."

As the local bishop, Bill Nebeker was the Mormon equivalent of a pastor and oversaw the Murtaugh ward as well as his own family's religious instructions. He often invited Matt to their Monday family nights. They were quite close and I appreciated his kindness. As the sky darkened, strings of multi-colored lights fired up around the awnings and amusement rides. As the approaching storm strengthened, I saw the colorful wires begin to twist in the wind, their lights banging against the buildings and dashing into the sky.

"Step right up. Six shots for five dollars," a vendor yelled as he ignored the upcoming storm. Mike and Matt headed to the booth, not to be deterred by the wind. They waited their turn and looked at giant stuffed animals they imagined they might win. The object was to shoot moving ducks, bobbing across the back wall. Just as they approached the counter, laden with bolted guns, rain drops hit us and a strong breeze brushed our faces. Then the saturated clouds opened in full force. Colorful wire lights untangled from the rides and fell in heaping messes. We crammed under an awning as a massive deluge poured from the sky. Merchant signs whipped off their places, awnings crippled in the gusts, and garbage cans blew over. Ten minutes later, the storm cleared. Nervous fairgoers headed to the exits, and we followed the pack, leaving the shooting vendor to gather his jumble of ducks and prizes. With our jacket collars pulled up, we proceeded to Mike's pickup. No matter the weather, we always attended the Twin Falls County Fair.

Chapter Thirty-Four

Boating, Camping, and Fishing

"Let's go to Swan Valley this weekend," Mike suggested. "The fishing is supposed to be great."

"Okay," I said. "Do you want to leave on Thursday or Friday?"

"We'll just take the weekend. We'll leave on Friday and be back on Sunday."

With aspen leaves turning a bright, golden yellow and shimmering in the crisp September wind that particular Friday, Mike hooked our Hyde drift boat and trailer to the back of his truck. He placed his motorbike inside the pickup while I filled our cooler with food and sodas, enough for three days of camping. By late morning we were on the road, heading toward Swan Valley, a stretch of land surrounded by the Targhee National Forest near the Wyoming border.

Twenty minutes before we reached our put-in spot on the South Fork of the Snake River, Mike unloaded his red motorbike at the take-out ramp. We continued another fifteen miles, following the contour of the river toward the east. In a public, paved parking lot filled with trucks and trailers, we transferred our camping and fishing gear to our trailered boat, and Mike expertly backed it down the ramp. He released the come-along and dropped the boat into the water. I found the "come-along" to be a funny name but it turned out to be a handy device farmers used to winch, or pull, all sorts of items. The ratchets kept the winch from unwinding and it was one of Mike's favorite tools.

I held the boat's attached rope and walked it toward the dock. Mike parked his pickup while I fastened the boat to a metal tie. Once we donned our life vests, Matt climbed into the pointed bow. Mike stepped into the middle seat, sat, and raised the oars. As soon as he had them in the oar locks, I undid the rope and leapt into the squared-off stern. We immediately hit waves and the boat swirled in the current. We moved at a fast pace around an eddy and down the river. Legions of trees lined the waterway, bending from the wind with their leaves twisting in the autumn breeze. Two hours after our departure, we arrived at a relatively flat clearing. When Mike had the boat nosed into shore, I hopped onto the bow, jumped out, and pulled it closer. Once on land we removed our vests and unloaded the boat, lugging supplies to a level site where earlier campers had already dug a fire pit.

Once they had gathered their fishing gear, Mike leaned over Matt and told him to use an Adams fly to attach to his leader. They pulled on waders, tightened their belts, and strapped on khaki vests. Plastic fly boxes bulged from several pockets and imitation flies dotted their sheepskin patches. Once they attached reels to their fly rods and threaded a floating line through the guides, they headed out. I stayed behind to set up camp. After years of fishing and photographing, establishing camp brought me the most joy. We had a green, two-man mountain tent that I staked to the ground. I added a fly sheet above the tent that kept rain from entering the air holes above the entrance. I placed padded ground mats and three sleeping bags inside. It was a tight squeeze but we managed.

Although I had never camped in anything but a rustic cabin when I lived on the East Coast, I learned to enjoy the outdoor life when I moved west. In spite of owning the Angler's business, I didn't crave fishing. When the fishing was hot, I had no qualms about putting down my rod and picking up my Nikon camera. Consequently, I was fortunate to have several of my photographs illustrate fishing magazines, major sport catalogs, and, of course, my calendars.

As the sun sank beneath the surrounding cliffs, Mike and Matt returned to the campsite. The smell of burning wood greeted the two anglers as they walked from the river. Three bulky logs circled the fire, blazing in the pit, and a pile of sticks rested nearby. Dinner was almost ready.

"I have bad news," I announced as they were putting away their tackle.

"Now what?" asked Mike as he turned to face me.

"Remember when we left home, our hamburgers were frozen?" I said.

"Yeah. So?"

"Well, I put them in the microwave to thaw," I answered. "They're still there."

"What? They're in the microwave?" Mike asked.

"Yup. But I have plenty of buns, ketchup, onions, and tomatoes," I said. "We'll make do for tonight. But I'll expect fish for tomorrow's dinner."

"You got it!" they said in unison as they sat on the surrounding logs and began to eat their meatless hamburgers. Thank goodness they were starving!

I joined them fishing the next day with my two-piece Cortland rod. While Mike rowed, I cast to the side, throwing slack in the line so the fly would drift naturally. Yes! A trout took the fly. It was a beautiful rainbow that gave a decent fight with two aerial displays. Once I had him close to the boat, Mike bent over the gunnel and easily removed the hook from its protruding lip. He supported the fish in the water and moved it back and forth, allowing water to pass through its red gills. Once the trout began to squirm, Mike let him go. We believed in "catch and release" unless we were going to have fish for dinner. We already had three in the boat and didn't need any more.

After an hour or so, I took the oars and rowed while Mike and Matt cast from each end of the boat. Rowing was one of my favorite activities while we camped. I could easily maneuver our boat down the South Fork unless we came to heavy white water. Then it was time for Mike to take the oars.

Before long, our mini vacation was over. I maneuvered the drift boat to the take-out spot, and the guys jumped to shore. Mike claimed his motorbike and rode it northeast to his parked pickup, fifteen miles away. Matt and I disassembled all our equipment and placed everything on the side of the public parking area. When Mike re-appeared, he backed his pickup and trailer down the ramp and attached the boat. Once the connections were completed and our belongings placed into his truck, we began the long drive back to Sky Ranch.

Chapter Thirty-Five

Harvest Time

As the harvest was in full bloom when we returned from Swan Valley, Mike rose exceptionally early the next day. While he showered, I put on a bathrobe and stumbled into the kitchen to make his breakfast. Once he sat at the round table in the dining nook, I handed him a cup of coffee and left to dress myself.

But harvest was only a process, not a conclusion as I had originally thought. It was not only a period to gather the crops, but also a time to prepare the ground for next year's produce. Trucks rambled from potato fields to the ranch scale, professional hay mowers cut alfalfa, and combines swathed the grain. Men and vehicles moved throughout the ranch in a fluid dance between man and machine, bringing empty trucks to the fields and returning with their beds crammed with crops. Everyone was busy, busy, busy.

It seemed like we were in the middle of a stirred-up ant hill. Even after dusk, Sky Ranch employees continued to harvest. I stood at our dining room window and watched the lights of gargantuan combines reaping a nearby grain field. The machines belched out thunderheads of brush while I continued to stare. Mike would soon be home, tired and craving dinner.

As the next morning warmed, I joined a half-dozen Mexican workers in front of the potato cellars. They were hired as clod pickers. Trucks pulled up and slowly tipped their beds backward, unloading the potatoes onto a rubberized, conveyer belt. Workers stood on a platform alongside the belt and pulled culls as the crop passed in front of them. These were potatoes too small to sell for human consumption. They would be saved and used for cattle feed. The workers also pulled vines, sticks, clods of dirt, or anything else that was not a good-sized Russet potato.

I stood beside a small woman, her head wrapped in a brightly colored scarf, as potatoes passed in front of us. Any non-potato items she missed, the next worker on the platform would catch and toss to the ground. As potatoes moved closer to the storage shed, the conveyer belt held only the desired tubers and the ground was littered with remains. The stored potatoes were destined for McDonald's, other franchises, and most Magic Valley grocery stores.

While I stood on the platform, I ignored the tiny clods and collected the largest spuds, the ones the size of small footballs. These I would send to friends and relatives back East, always adding a note stating that they were Idaho's "average" potatoes. Standing on the platform as potatoes moved past me on the conveyer belt, I felt like Lucille Ball at the candy factory. I couldn't seem to keep up as they moved so quickly in front of me. One time the machine stopped and I practically fell over. My body had been swaying right along with the moving potatoes. The workers laughed at my inexperience, and I laughed right along with them.

Harvest time was always hectic. During one extremely busy period, our neighbor, Dean Moss, suffered a major health issue and could no longer work. I was surprised by the local reaction. Even though it was in the middle of harvest, all the nearby farmers put in extra work, men, and machinery to assist the Moss family. In record time, their crops were gathered. Although sadness consumed the area, the family didn't lose their income. While the relatives suffered, neighbors brought food and comfort to the home. I went over with a plate of brownies and visited with Marsha, revisiting funny stories we shared throughout our years together. Matt played in the living room with her children while Mike helped with their harvest. Dust consumed the land as neighbors worked from row to row and from field to field. Clouds collected into a squall, but no rain appeared. Before long darkness had snagged them. I saw lights from the combines slicing through the fields as they continued to harvest. A red moon crept over the horizon. The men turned off their machines and left the fields. Work would continue at the Moss ranch the following day.

Living in rural America, I discovered that it was almost an unwritten rule: when someone was hurt, everyone pitched in to assist. We were glad to help. And we knew if we had experienced a tragedy, our neighbors would have been there for us, too. It's just the way it was.

* * *

October slipped into November and the smell of burning farmlands seeped into the sky. Clouds of smoke rose from the fields as the days grew shorter, the air turning cold and crisp. This was the time to take Matt and a few of his friends into nearby fields now reaped of their crops. I drove onto a harvested onion field about noon, making sure to drive parallel to the rows of brown dirt to get to the middle of the field. The boys scrambled from my new Jeep SUV and pulled black bags from its trunk. In jeans and flannel shirts, they scattered like autumn leaves, two to a bag, running to pick as many onions as possible. The big machines couldn't reap all the onions, so we'd pick the leftovers. It was called gleaning. Once the black bags bulged, the boys dragged them back to my car. In the heat of the day, the trunk smelled like French onion soup. From there we went to pea fields and then on to our potato fields. Gleaning was like a gigantic Easter egg hunt. But instead of colorful eggs, we collected potatoes, pea pods, and onions. And the children learned the value of crops, whether they had been harvested or left in the field for us to gather.

Chapter Thirty-Six

Christmas and New Year's Activities

Weeks passed and the days began receding, becoming shorter and darker. Soon late autumn winds swept across the valley, taking with them its beautiful fall colors. The nights were cold as I stood on our back steps and looked at the sky. The light of the moon, bright on long grasses that ruffled in the rising breeze, brought a feeling of mystery and wonder. The nearby foothill colors had faded, and light flurries heralded the holiday season.

Throughout the growing year, the ranch had looked like a patchwork quilt: green potato leaves, gold grain grasses, olive alfalfa, and lime-colored peas. The fields had changed their subtle shades of color with each passing month. Finally, the snows came, and the ranch converted to a white blanket, stretching across the barren valley.

Strong storms swept from the Sawtooth Mountains, and colored lights blazed in downtown Twin Falls. The streets, arched with bare tree limbs, were crusted with ice. Groups of shoppers bustled about in arctic gear and strolled along winding sidewalks, their faces bathed in the intermittent lights of shop windows. Mike, Matt, and I wore down jackets and wool caps and were swept along Main

Street by a chilly wind at our backs. We walked past Sav-Mor Drug and Rudy's Price Hardware, their windows decorated with red ribbons and green wreaths while shoppers waved and wished us "Merry Christmas" as they hurried through the chilly night. Once my gifts were bought and wrapped at The Paris, a high-end women's store, I returned the pink credit card to my purse. The three of us convened under the old-fashioned streetlamp in front of Roper's Clothing Store, its windows now glowing against the gathering dusk.

From Main Street, Mike drove us to Rock Creek Restaurant on Addison for an early dinner. A stained-glass window, depicting a jumping trout, greeted us in the foyer. Mounted fish, ducks, and deer enhanced the inside wood-paneled walls. At the far end stood a large, brightly decorated Christmas tree. White lights draped across the salad bar with its ever-present, homemade chowder.

"Hi, Stan," Mike called as we entered.

"Hi, guys," Stan hailed us from behind the bar. "You're awfully early, aren't you?"

"Yup, we're in town to shop and buy a Christmas tree."

"Glad you're joining us. We have a steak special today," Stan said with a wide smile.

We hung our jackets on hooks next to a booth and slid into our seats. Matt wedged his back into the corner and looked at the menu. Mike and I decided on Angus beef and mashed potatoes.

"We'll have the special, medium rare. Matt wants chicken fingers," Mike called to Stan. "And I'd like a Coors. Bobbi wants a glass of merlot, and Matt wants hot chocolate."

"Coming up," Stan answered and turned to place our order with his cook. A hint of baking bread drifted in from the kitchen, and our stomachs started to growl. Stan retrieved the drinks and brought them to our table, wiping his hands on his white apron. After we'd settled in and chatted for a few minutes, Mike rose and headed to the salad bar. Matt and I followed, making sure we each had a cup of corn chowder along with an assortment of salad vegetables and some homemade rolls.

Later, the three of us drove to the tree lot in the Albertson grocery store's parking lot and meandered between rows of upright spruces. Leaning against an old travel trailer parked at the five-points corner of Blue Lakes Boulevard were layers of wrapped spruce trees. After inspecting several, we found a full one, about eight feet tall. A small man with a black mustache, wearing padded overalls and a black wool cap, approached us. He told Mike the price and I asked for the leftover branches to create a wreath and to decorate our fireplace mantel.

"They're free," he said. "Here, let me help you get that tree into your truck."

"Thanks," I said as the men heaved the tree into Mike's pickup.

As soon as we arrived home, we pulled the spruce from the back of the truck and placed it on the cement pad behind the garage. Matt and I held the tree while Mike cut several inches from its base. From there, we carried the heavy tree and wrestled it through our French doors while Matt ran to retrieve a heavy towel from our upstairs bathroom. Once inside, I put down the towel and added a tree stand. While I was on my hands and knees, Mike lifted the tree, high enough for me to place the trunk into the stand. I crawled underneath the bottom branches and fastened long screws into the thick trunk. While I filled the base with water, Mike and Matt untangled extensive strands of colored lights. The three of us wrapped the cords in and out of the branches until the tree was completely covered. Mike plugged the string into a socket and the tree lit up. The sweet scent of sap soon filled the air. We hung an assortment of familiar ornaments, and then came time to tackle the dreaded tinsel.

"Only one strand at a time. No bunching and no throwing," Mike commanded. Single tinsel strands were a Wolverton tradition. He worked on the top of the tree, I took the middle branches, and

Matt scrambled around the bottom. We stood back and inspected our creation. The tree looked beautiful. Christmas music played on our record player and snow started to fall.

Everything seemed picture perfect, but I wanted more. I wanted Christ in Christmas. I thought the season had become too commercial. Growing up in a small town in Connecticut, I remembered a community service that many had attended on Christmas Eve. I thought we could duplicate the same thing in Murtaugh. As a member of the Methodist Church, I asked the women's guild if we could have a Christmas Eve service. The ladies would provide punch and cookies, I would organize the event, and the whole town would be invited.

"What do you mean by the whole town?" one of the women asked.

"Well, everyone. No matter their age or their religion," I said.

"The Mormons and the Catholics won't come. They have their own services," she stated.

"Let's try. I just need your permission," I said.

They voted and approved the church's first Christmas Eve service. Once they decided to move forward, they did so at lightning speed. The women talked about which cookies to make, what type of punch, and whether to have coffee. It was a rush job but they were up to it. I just had to create and manage it.

"Hey, Kay," I called over the telephone to the Nebekers' that evening. "Can you help? I'm organizing a community Christmas Eve event, and I need some singers. Do you have a men's choir at your ward?"

"Yes. It's fabulous."

Once she gave me the name of the choir leader, I called him and asked if his group would be interested in singing on Christmas Eve at the Methodist Church, not for just the church but for the whole community. He told me he'd get back to me but he was pretty sure they'd do it. Next, I called three Murtaugh leaders, two women and a man, and asked if they'd join in the celebration. They all said, "Yes." And Dale Metzger, the Methodist pastor, volunteered to obtain candles.

"Anything you need, just ask me," he said. "I'll be out of town that week but I'll help any way I can."

My good friend and postal employee, Jane Toupin, volunteered to play the piano. There would be no sermon, just three adults reading verses from the Gospel of Luke, and the congregation would sing a half dozen Christmas carols. Surprisingly, on that first Christmas Eve service at the Methodist Church, many community members supported the event. They entered the warm vestibule and before long, we had over a hundred people, from infants to elderly seniors. Considering we usually had twenty or so people attend the regular Sunday services, it amazed me that so many came that night. Not only was it a last-minute event, but it started snowing in the afternoon. It was definitely a season of miracles. The familiar carols, sincere and sweet, filled the old-fashioned church.

At the end of the thirty-minute service, everyone walked to the front of the church, collected a candle, and lit it from the advent display. Then they returned to form a giant circle around the perimeter of the nave, facing the now-empty pews. Kay stood near the backdoors and turned off the lights. What a sight. All these people stood with their backs against the church walls, their lit candles reflecting the rosy glows in their faces. The last song was "Silent Night." In soft, flowing voices, the gathering sang the three verses. I wiped a tear from my cheek and marveled at the beautiful occasion, the sacred joy of the Nativity. As the town folk left the church, they took their candles and placed them in small snow banks surrounding the church. Mike, Matt, and I turned to face the white edifice surrounded by candles, sparkling in the snow. It could have been a postcard photo. People were chatting, hugging family and friends, and thoroughly enjoying the moment. For me, Christ had come back into Christmas.

* * *

After the spiritual celebration, Mike and I prepared our house for a New Year's gathering, asking fishing friends from Southern Idaho to join us. We usually had six or seven couples coming for a New Year's Eve dinner, an overnight stay, and a full breakfast the following morning. Each couple brought and served one course while I managed the vegetables and entrée.

A few days before the group's arrival, snow had spun a cocoon of white all around Sky Ranch. Sunlight sparkled on ice crystals hanging from our trees; it looked like a fairyland. Mike and I extended the dining room table and covered it with a giant, white-embroidered cloth from Italy. We placed chairs around the table, and I arranged the formal dinner place settings with separate wine glasses for each of the six courses.

Out came the crystal, china, linen, and silver. This was my time to show off. The wine glasses sparkled, the silver glistened, and the house flashed with Christmas finery. It usually took me two days to complete the presentation. An hour before the guests arrived, Mike and I began dressing. I wrapped my brown hair into a French roll and added a glittery comb along its edge. I added a sequined red blouse and black gaucho pants along with strappy high heels.

"Mike, can you zip up my top?" I asked.

"Sure. If you'll sew on a button. It just popped off," he said. "I don't want to get undressed, so can you do it while I'm wearing my shirt and sweater?"

"Yup. Let me get some thread."

After he zipped the back of my blouse, I turned and threaded a needle and asked Mike to hold the shirt's cuff. He grasped it between his right thumb and forefinger and pulled the shirt's sleeve away from his navy sweater. I tied a knot in the end of the white thread and started to weave the needle in and out of the button holes. When I struck the button, I firmly shoved the needle and pulled it through to the other side.

"Christ!" Mike shouted as he flinched and jerked away from me.

He still held the shirt cuff, but I could see the needle had gone through his thumb. I had the needle in my hand and twitched it up and down as his hand moved in response, performing like a puppet.

"That's not funny!" Mike roared. But I was already howling with laughter.

"Okay. Wait a minute," I said, trying to stifle my laughter. I grabbed some scissors and cut the thread and rethreaded the needle. This time he held the cuff with the left hand, and the button was sewn correctly with no drama. He accompanied me in silence as we walked down the hall to the kitchen, scowling at my stifled snickering.

Guests started arriving as soon as dusk began to settle in the valley. The snow covering our lawn looked like a blanket of white gauze. Our porch light illuminated the front entrance and colorful lights decorated the bushes below the brick gable. Matt had been dropped off at Kay's house earlier in the day. She had games planned with her seven children, and Matt knew he'd have his own evening of fun.

Once everyone had drinks and played a little pool, we approached the dining room and sat for the formal dinner that began each New Year's Eve at eight o'clock. Close to midnight we were on our last course. I turned on the television, and we stood to see the ball drop at New York City's Times Square. Flushed with champagne, we stepped outside the living room French doors. The bitter cold night showed stars like icy pinholes in the dark sky. Looking above we saw Orion's Belt, the Big Dipper, and the Milky Way. To be in a beautiful setting with wonderful friends was just the perfect way to start the new year, or so we thought.

Chapter Thirty-Seven

Deaths All Around

January began with sadness. Mike's mother, Margaret, had suffered a stroke a few years earlier and eventually succumbed to her medical condition. Months afterward, my eighty-six-year-old father had a massive heart attack and died.

Twelve days later when we were sound asleep, our bedside telephone rang. Mike awoke slowly and fumbled in the darkness to light the lamp by his side of the bed. I looked at our illuminated clock. Who would call us at five in the morning? Mike reached over and answered the phone and then walked into the kitchen. When he came back to bed, I asked, "What was that all about?"

"Your mother just died."

"What do you mean? My father just died."

"No. Your mother just died. The sheriff called from Florida. I wrote his name and number on a pad in the kitchen," Mike answered. "Give him a call. He'll tell you details." As I rose and rushed to the kitchen, Mike asked me to close the door so he could go back to sleep. It was too early for him to get up.

How could my mother die when my father had just died? Although my parents were in their eighties, they had been in relatively good health. I telephoned the sheriff and heard that my mother had died early that morning from complications due to her recent carotid artery surgery. My older sister had been with her for the past ten days but was now in Wyoming, researching a dude ranch for her latest book. There was no way for me to reach her as she was in the mountains on horseback. I spent the next few hours calling relatives and trying to understand, sobbing as I talked on the phone. Losing both parents in such a short time was heart wrenching. In spite of many friends consoling me, I became overwhelmed with grief. There was no one to hold me and no one to share my pain. I felt like a solitary cloud, isolated and lonely.

Reeling from the blow of my parents' deaths, I went about my daily chores like a robot: making meals, tending to household chores, and working in the office. I couldn't seem to shake my sorrow. Eventually, as weeks passed, I slowly began to feel better. Not happy, but better.

One day a few weeks later, I saw Smoky lying on his side in the dirt near his barn. He didn't appear to be moving. Not even a twitch when a large bird landed beside his body. I was surprised to see him so still.

"I'm going to check on Smoky," I told my staff as I glanced out the office window.

Not overly concerned, I moseyed toward the motionless horse, calling his name as I advanced. Polly grazed far out in the field as I approached her buddy. Smoky raised his head, coughed several times, extended his forelegs, and slowly rose to his feet. He shook dust from his body and coughed a few more times. His head hung down, almost touching the ground. I rubbed his body and the side of his face. Yellow mucous dripped from his nose.

Hurrying back to the house, I entered the kitchen and telephoned our veterinarian.

"Dr. Monroe? This is Bobbi," I said. "Smoky's sick. Can you come over?"

"Sure. I'll be there after lunch," he answered after a short pause.

When he and Ruth pulled into our driveway, I had figured out the problem. A few months earlier, Mike had had one of his men bring a ton bale of hay to the area between the barn and the shed. A wood pallet designated the selected spot. One horse cannot eat that much alfalfa before mildew starts to infect the hay. I inspected the giant bale and saw white shreds of fungus. With all his coughing, I surmised that Smoky had developed a lung infection. Dr. Monroe confirmed my suspicion. He gave Smoky an antibiotic shot and handed me some large, white pills to give him over the next five days.

"Hey, Mike. Can you have one of the guys bring another bale of hay? This one's bad. Clear."

"Will do. Clear."

Later that afternoon, I checked Smoky. He seemed so much better. He was up and grazing with Polly, his shadow, right beside him. The sun blazed over the prairie and I saw more colors of olive and lime. Spring had arrived and thirty-year-old Smoky was finally healthy.

Several mornings later, I was again upstairs in the office. As I turned my attention to orders and deadlines, telephones began humming. Karen called me to her desk to approve an invoice. As I stood beside her, I looked out the office window and saw Smoky lying on the ground again.

"Now what's happening?" I said as I sprinted down the back stairs and out the garage door.

Smoky lay about thirty feet from the barn, his head resting in the dirt. Spasms assaulted his lungs, and in between, he lay absolutely still. There was no movement until he coughed again. His whole body shook with each cough. His nose filled with yellow mucous and his breathing was abnormally labored. I couldn't get him to move, not even to raise his head.

"What's up, Smoky? You were doing so well," I said as I knelt next to his face. "Come on, Smoky. Try to get up."

He didn't move. I sat on the ground and placed his head on my lap. I couldn't believe its weight. As I rubbed his face and brushed my hand down his forelock, he opened his eyes and looked at me.

"Come on, Smoky," I murmured. "You can do it. Please try."

He stirred his limp muzzle and let out a tiny whimper. I tried to lift his head but it was impossible. It was too heavy. I cradled it in my lap, stroking his face and wiping his nose. I leaned forward and bent my shoulders over his head and cried uncontrollably. Sobs engulfed my body. He was actually dying. I knew it. I felt it.

Smoky parted his quivering lips and let out a long sigh. Then nothing. A single tear rolled down his cheek. He stared unseeing and I knew he was gone. I sat on the ground, holding his head. Convulsing sobs consumed me. My heart was breaking. We were a team and he had died. I don't know how long I sat in the dirt with his head in my lap. I couldn't seem to leave.

"Come on, Bobbi. Get up. Mario'll take care of him," Mike said as he extended his hand and pulled me toward him. I turned and took one last look at my horse lying in the dirt. We walked to the house with his arm around my shoulder and I departed for our bedroom. I didn't make lunch and Mike never asked me to. Tears filled my eyes as I collapsed on the bed. I couldn't believe my Smoky was dead. Totally bereft, I sobbed into my pillow.

The next day was unbelievably miserable. Yet, I fulfilled my obligations and made breakfast for Mike and Matt. I couldn't eat and performed my responsibilities with only the minimum of attention. Mike didn't seem to grasp my sadness.

"What's for dinner?" he asked as he stood in the dining room, looking in the mirror and moving his baseball cap to just the right position. He was going to check on his employees. Once he left, I walked to the barn to feed Polly. She wasn't there and she wasn't in the pasture. Polly had vanished. How could a huge pig vanish into thin air? After informing Mike and Matt of her disappearance, they climbed into his pickup and searched our nearby roads. She was nowhere to be found. When they returned home, Mike called his staff.

"Keep a look out for Polly," he said over the ranch radio. "She's gone missing. Clear."

"Will do. Clear."

Another hour later we heard chatter on the radio. We couldn't understand the words and then they came through loud and clear.

"We found her. She's down in the gravel pit. Do you want us to bring her home? Clear."

"Yup. We'll help. Clear."

The three of us piled into Mike's pickup and drove toward the gravel pit. Mario had already enticed Polly out of the rocky hollow and had her on the side of the road. One of the ranch pickups had been parked nearby and had its tailgate open with a wooden plank slanted down from its back. Mario jumped into the truck's bed with a bowl of pig food. Danny, Mike, and Melvin pushed her up the plank while I photographed. Once she was completely inside the rails, Mike slammed the tailgate shut.

"Now, that's one big pig," Mike said as he wiped his brow and brushed pig dirt from the front of his shirt and jeans.

"Before you bring her home, would you take her to the scales?" I asked. "I want to see how much she weighs."

Once Mario brought Polly back to her barn and reweighed his truck, we heard she weighed six hundred and twenty pounds. Yup. She was one big pig.

We weren't in the house more than a few minutes when the telephone rang. Kathy Adams, Larry's wife, had a question for me.

"Was that a black cow you were bringing home?" she asked.

"No, why?"

"Well, you have a black dog, black cat, black horse, and black pig. I thought you were getting a black cow."

"Nope. That's Polly," I answered. "She escaped this morning."

"Yikes! She's enormous," she exclaimed.

"Yeah. We just weighed her. She's over six hundred pounds!"

"That's like a steer," Kathy said.

"And I have some sad news, Kathy. Smoky died yesterday."

"Oh, I'm so sorry. I'll let Larry know," she said. "What happened?"

In between tears, I told her the story of Smoky getting into bad hay and not being able to clear his lungs. He had been such a special horse to me, more like a pet than a working horse.

"I think that's why Polly left," I continued. "I think she was looking for Smoky. From the time Polly was a month old, she followed Smoky everywhere. I don't think she realized Smoky died."

The days went by as usual until a week later when Polly disappeared again. I walked out to the road and noticed sheep tracks in the dirt. A flock must have been driven past our house earlier that morning. They would have been on their annual pilgrimage from the foothills to nearby pastures and eventually to market. Knowing the flock would be herded by men on horses, I walked back to the house.

"Hey, Matt." I called up to his bedroom. "Let's go find Polly."

He rambled down the stairs and walked into the kitchen. After I told him that Polly had escaped again, he joined me in my Jeep. We followed sheep tracks three miles to the west.

"Have you seen a pig around here?" I asked one of the cowboys when I caught up to the flock.

"Yup. We did," he said. "It was the funniest thing. We passed this house and a huge pig came running out after us. I've never seen a pig that big!"

"Did you see where she went?"

"Yup. She turned north at a horse corral."

I knew exactly where he meant and reversed my direction and drove back toward our neighbor's ranch. There in the small enclosure were three horses. Flakes of hay had been strewn on the ground near the fence. Polly was on one side of the horizontal logs and the horses were on the other side. I could tell she was delighted as she chewed the hay alongside her new best friends, her tail twitching happily back and forth. The nearby horses were anything but happy. They flattened their ears, curled their lips, and bared their teeth. When Mario appeared this time, Polly easily mounted the wood plank and rode in the back of the truck to our barn. Once she was settled with some extra food, I returned to the house and began our dinner. With my hands buried in a ceramic bowl of meatloaf, I told Mike about Polly's latest adventure.

After our meal Matt joined me in the barn. The two of us cleaned out Polly's stall and added more straw. I picked up a five-gallon bucket and began to fill it with water. Without warning, Polly charged Matt, knocking him to the ground. Not even think-ing, I reacted immediately and spun around, pouring the bucket of water over Polly's head.

"No! No! No!" I shouted.

Polly ran from the barn and I chased right after her, swinging the bucket, trying to hit her. Even for a gigantic pig, she was too fast. I returned to the barn, out of breath and very angry. Matt stood at the entrance, brushing off his shirt and jeans.

"Why'd she do that?" he asked.

"Who knows?"

I could only surmise that Polly thought Matt was too close to her food. Polly outweighed Matt by over five hundred pounds. There was no contest. I also remembered reading in the newspaper the first year I arrived in Idaho a story about an elderly man falling into a pen and being eaten by his pigs.

"Matt, you'd better go back to the house," I cautioned. "I don't think you should be here. Without Smoky, she's become too dangerous."

He left and I finished my chores. When Polly came back in the barn, I brushed her off and gave her a bucket of food. As I straight-ened items in the barn, she gobbled her slop and sauntered into her stall. Usually, she'd lie on her side and I'd sit beside her rubbing her belly. Tonight, no longer comfortable with the ritual, I stayed outside her stall and leaned on a rail, making soothing sounds, try-ing to coach her to relax. She looked at me with piercing eyes and growled.

What the hell? I thought.

I didn't even know pigs could growl but growl she did. If it weren't safe for Matt to be in the barn, it certainly wasn't safe for me. And now I was there by myself. If anything happened, no one would know until it was too late. In retrospect, I should have shown Smoky's dead body to Polly. I was so overcome by my own grief, I never thought of Polly and what she might have been feeling. She didn't know Smoky had died. Because she couldn't be with Smoky or with other horses, she had become mean. I made a tough decision, unpleasant but absolutely necessary. Since Mario fed Polly each morning, he had a good relationship with her. I telephoned him and asked if he'd take care of her.

"You have to use a gun. And only one shot. Can you do it?" I asked.

"I'll do it. When?"

"We're leaving this weekend for the Little Farm. Can you do it then?"

"Okay. I'll handle it."

I didn't want to know any details, but I did hear that his family and friends joined him for what they said was the best pig roast ever. His wife, Louisa, made her special tamales, wrapped in corn husks, while others brought beer and extra food. Mario couldn't help it and told me the pork was lean and absolutely delicious.

Chapter Thirty-Eight

Selling Sky Ranch

A t the end of harvest that year, prices for ranch crops were at an all-time high. Don and Mike decided it would be the perfect time to sell. They had been contemplating it for several years as no other family members wanted to take on the ranch enterprise.

Although I loved living in rural Idaho, I didn't care whether they sold Sky Ranch or not. Except as a curious bystander, I was actually never involved in the operation. My business was the Angler's Company, and since Mike and I owned the Little Farm, I had all the farming I could handle.

Once word got out about a possible sale, two interested buyers competed for Sky Ranch. The selected owner asked the brothers to help with the transition and to stay on for one year. During that time, our family moved to Buhl and Mike commuted to Murtaugh. After we settled into our new house on Country Club Drive, I sold the Angler's business. Our life styles changed dramatically. Mike and I took up golf and I no longer made three meals a day. Yet we still kept our hands in farming. Almost every weekend we'd drive the short distance down River Road to continue improving the Little Farm.

When the year-long transition at Sky Ranch was fulfilled, any item the new owner did not buy was consigned to Musser Brothers

Auctioneers. Representatives from most Western states and three Canadian provinces attended the farm sale that March. Men in thick, heavy clothes stood in a semi-circle surrounding the rear of the ranch headquarters. Vehicle after vehicle rolled in front of the men. Trucks, tractors, backhoes, loaders, and combines. I heard Randy Musser, the auctioneer, on a portable microphone calling out bids.

"Sold," he shouted over the crowd. And another vehicle rolled to the front of the line. Voices increased as more farmers became involved in the bidding. Dark clouds formed and snowflakes started to fall. A slight wind drifted from the mountains and freezing mud surrounded the cluster of men. They stomped their feet and rubbed their hands as the food truck passed out hot coffee. After the heavy equipment sold, Musser moved with his microphone to the assorted attachments scattered on the ground: blades, shovels, brush cutters, buckets, scrapers, and harrows. I photographed from the top of a tractor cab, fascinated by the process. Soon the auctioneering company gathered the slew of farmers and they moved inside. Musser went from room to room. Everything sold. From Motorola radios to nuts and bolts. There was nothing left when he had finished.

It was a sad day. Dark and cold. Ranch employees and business acquaintances shook hands or hugged Don and Mike as they said, "Goodbye." It was a passing of their life's work, the end of an era.

* * *

Alone in the afternoon gloom a few days later, I sat in my car and looked at the section of Idaho I had grown to love. During the past fourteen years, all kinds of unforeseen events surprised me as I ventured into that unfamiliar territory. I became a wide-eyed witness to what was actually required to run a corporate farm. And I learned to live miles from civilization, to run a home business, and to satisfy my ranching husband's needs.

I came to Idaho having a solid foundation of hard work, loyalty, and honesty. When I moved to the desolate ranch, I welcomed a new life of challenges and subtle unknowns. I'd stand on our back steps and look at a night sky abundant with bright stars and feel the cold winds raging from the distant mountains. I always thrilled to the pulsating sounds of huge farm machines working in nearby fields and loved the smell of pungent sage as I walked across the prairie. It was with sadness that I remembered those magical moments. Tears rolled down my cheeks and I bowed my head and gave a silent prayer of thanks for the insights, compassion, and lessons I had learned. Snow softly fell and drifted over the vast valley as I drove away from Sky Ranch for the last time.

Epilogue

Once we moved to Buhl, Mike's and my differences became more pronounced. We began to go our separate ways. Although we pursued marriage counseling, we couldn't seem to make amends. Before long, our slight crack became a chasm; we divorced and sold the Little Farm. We had had over twenty years together, all with many challenges and yet also with so many rewards.

Mike now lives in the Boise area and spends his winters in Arizona. I married a high school boyfriend that I reconnected with at a class reunion and retired to Tennessee. Matt joined the United States Coast Guard after college and lives on the East Coast. I continue to visit friends and relatives in Idaho every couple of years. My life is different. But life goes on.

Glossary

Angus: A breed of cattle that originated in Angus, Scotland. They are solid black.

Belly Dump: A V-shaped trailer that empties its load from the bottom. It can carry sand and rocks, as well as crops, and has the advantage of laying materials in rows.

BLM: Bureau of Land Management, a Federal program that manages land throughout the Western states.

Boar: A fully-grown, uncastrated, male pig.

Borrow Pit: A trench dug on the side of a road to provide drainage and furnish road fill by "borrowing" dirt from the roadside ditch. Sometimes called a "barrel" pit because of the wheel barrels that were used to remove the dirt from the roadside.

Bovine: Cow.

Burn Barrel: Fifty-gallon drum that people use to burn household items outside.

Cattle Guard: A pole-like contraction that lies in an indentation on a road, keeping livestock from crossing. It acts as a fence without having to open or close a gate. It's usually painted bright orange or yellow.

Cattle Prod: A handheld device with electrodes used to move livestock by poking them with a high-voltage, low-current electric shock. Also known as a hot shot.

Chaff: Inedible parts of grain or corn . . . the stalks or husks.

Charolais: A white or cream-colored breed of cattle that originated in France. They are known for their gentle demeanor.

Combine: A field machine that combines three aspects of harvesting: reaping, threshing, and winnowing.

Come-along Winch: Farmers used the device to winch, or pull, all sorts of items. The ratchets kept the winch from unwinding.

Culls: Small or damaged potatoes. They are often used to supplement cattle feed.

Dally: To wrap a rope around a saddle horn after lassoing a cow.

Ear Tags: Plastic identification tags fastened to a cow's ear. Each tag displays a number detailing the history of that individual cow. It replaced branding.

Fallow: A section of uncultivated land. Instead of using a rotating crop, farmers might leave a field fallow, so it can rejuvenate.

Harrowing: Using tines, the harrow loosens the soil, preparing it for planting.

Header: A cowboy and his horse are both considered headers. The cowboy ropes the head of a calf and his horse backs up once the animal has been lassoed.

Heeler: A cowboy and his horse are both considered heelers. The cowboy ropes the hind leg or both legs of a calf and his horse backs up once the animal has been lassoed.

Heifer: A young, female cow.

Hereford: A mostly brown breed of cattle with white markings. It originated in Hereford, England.

Mother-cow: A mother cow with calf, often called a "MC."

NRA: National Rifle Association.

Reaping: Cutting a crop.

Siphon Tube: An S-shaped tube that transports water from a canal to a row between planted crops.

Sow: A fully-grown, female pig.

Squeeze Chute: A metal container, often portable, that controls an animal by squeezing its sides and locking its head.

Threshing: Separating grain from the inedible chaff or stalk.

Till: To till the soil is to turn it over by plowing and harrowing, preparing it for planting.

Tiller: A lever, or arm, attached to a boat motor to steer the vehicle in a certain direction.

Tines: The pointed, sharp, metal rakes used to turn over soil to prepare it for planting.

Ward: A Mormon meetinghouse.

Windrow: A long line of cut hay, straw, or grain that is laid out to dry in the wind.

Winnowing: Removing the inedible chaff or stalk from the grain.

Yorkshire: A white- or pink-colored breed of pig that originated in England. They have erect ears, can live past ten years of age, and weigh over seven hundred pounds.

Maps

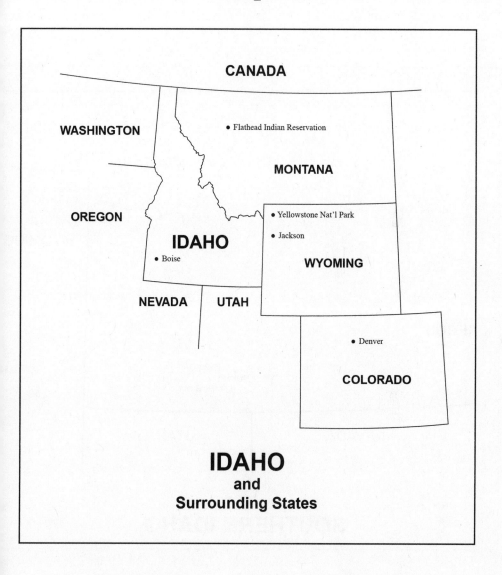

IDAHO
and
Surrounding States

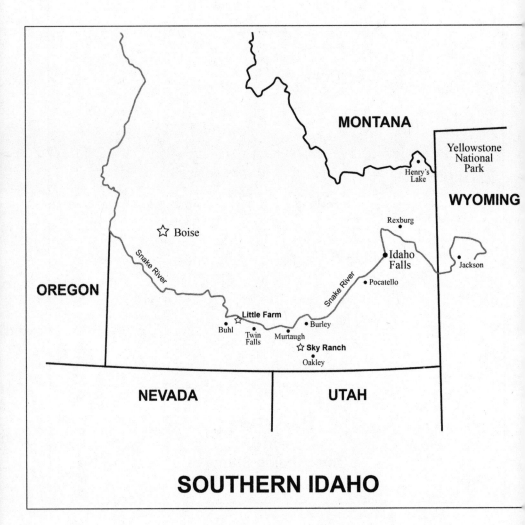

SOUTHERN IDAHO

About the Author

After graduating in 1968 from the University of California, Berkeley, I became an international flight attendant. For six years I took troops to Europe and the Orient, including Vietnam during the height of the war.

In 1975 I began the Angler's Calendar and Catalog Company. A few years later, I moved to Idaho and worked for the *Times News* as an advertising representative for the Twin Falls newspaper while continuing to expand the Angler's business. In 1985 I was presented with the Arnold Gingrich Writers award from the Fly Fishers International organization, and in 1993, my company won Exporter of the Year for the State of Idaho. As a twenty-year member of the Outdoor Writers Association of America, a three-year board member of the Fly Tackle Trade Association, and a nine-year member of The Nature Conservancy of Idaho, I connected with sports-oriented people throughout the world and my Angler's business continued to expand. I sold both the Angler's Calendar and Catalog companies in 1995.

When I remarried and moved to Tennessee, I began to write books. The Authors Guild of Tennessee voted me in as president in 2015. See authorsguildoftn.org.

Acknowledgments

To all my Idaho friends, relatives, and employees who helped me on my journey from being a city single to a country wife and mother, I can't thank them enough. It was through their advice and kindness that I survived challenging issues while living on a remote ranch in South Central Idaho. And it was through their encouragement that I continued to learn, absorb, and thrive.

My Angler's Calendar employees included Diana Breeding, Karen Brown, Alanna Dore, Richard Fuehrer, Pam Grimm, Cathy Humphries, Sharon Kimber, Christie Tipton, and Jane Toupin. Alanna Dore was my first and only employee in 1975 when I started the business in Berkeley, California. Hal Janssen, a fishing friend and contract artist, illustrated fishing flies and wildlife that I overlaid on each monthly calendar grid. The company continued to grow and eventually produced and sold fourteen different sport calendars. I had four employees working in the space above the house garage out at Sky Ranch who annually contracted 1,600 sport and book stores throughout the world. We had over 200 companies who paid to be listed at the back of the fishing calendars (fly, saltwater, and bass). Listed were fishing shops, guides, lodges, manufacturers, schools, and publications. The other segment of the business was the Angler's Catalog Company. My firm produced a forty-eight-page color catalog, depicting high-end fly-fishing gifts, that was mailed to 200,000 households. Both businesses sold in 1995.

My special thanks to Nick Lyons, original owner of Lyons Press and former monthly columnist for *Fly Fisherman Magazine*. He mentored me for over forty years, first with the Angler's Calendar Company and then with my latest writing endeavors. He nudged me to write a better book by suggesting I convey the stark contrasts from my life on the East Coast to that of living at Sky Ranch.

I am forever grateful to my writing and reading friends who proofed *Sky Ranch*, from its earliest conception to its final form. They gave me the gift of critical feedback and positive suggestions. They are: Cheryl Adamkiewicz, Patricia Crumpler, Jody Dyer, Kathy Economy, Ginger Rogers, and Lynn Toettcher. Kay Nebeker and Susan Kite proofed the Mormon chapter; my sister, Ginny Clemens, proofed the horse chapters; and Don, Georgina, Matt, and Mike Wolverton as well as Mychel Matthews helped me with correct cattle and farming terms. And a special thanks to my loving husband, Larry Chapman, for his kindness and understanding during the stressful time of writing my book about ranch life.

How lucky I am to have Skyhorse Publishing take a chance on me. Tony Lyons, president, and Bill Wolfsthal, vice president, chose to accept my book. Associate Editor Caroline Russomanno patiently walked me through the process of a manuscript submission with understanding and finesse. I couldn't ask for a better team. Bob Ballard of Working-Class Publishing tweaked their original cover design and made it the best one ever. He also reconfigured the two maps I drew. Thanks to them all.

My Fly Fishing Friends:
Anne and Darwin Atkins, Bonnie and Ray Beadle, Silvio Calabi, Vicki and Roger Cantlon, Del Carraway, Betty and Ron Cordes, Nick Curcione, Laurel and Jim Dixon, Janet and Marty Downey, Chris and Mike Fong, Penny and Mike Glenn, Sue and A.J. Hand, Jane and Hal Janssen, Keith Kiler, Lefty Kreh, Fanny and Mel Krieger, Shirley and Buddy Macatee, Maggie and Ron McMillan,

Jackie and Craig Matthews, Maggie Merriman, Charlotte and Larry Miller, Gerri and Pat Moran, Darlene and Ed Rice, Barbara and Alan Roehrer, Carol Roseberry, Ernie Schwiebert, and Joan Wulff. We camped and fished together from Alaska to Christmas Island in the Pacific, from Nova Scotia to the Caribbean, from Idaho to Argentina, and many spots in between. But mainly we fished and camped in the Rocky Mountains. We were all active in Fly Fishers International.

Friday Night Bridge Group:
Over our many years together, we shared outdoor adventures of skiing, fishing, and camping; pregnancies and births; children's weddings; and the sadness of losing our parents. We initially came together through the Twin Falls Welcome Wagon association and stayed friends for over thirty years. The original Friday Night Bridge group consisted of Mary Lu and Gordy Barry, Janet and Paul Beeks, Sue and Bill Carver, Mary and Tom Courtney, Lois and Mike Cowan, Cathie and Dave Dellett, Cora Lee and George Detweiler, Jan and Dick Greenwood, Peggy and Ray Hackley, Pat and Fred Harder, Debby and Ron Miciak, Dottie and Bob Miller, Sherry and John Ritchie, Nancy and Doug Strand, Dar and Tom Wagner.

www.booksbybobbi.com
www.authorsguildoftn.org